4th International Workshop on New Computational Methods for Inverse Problems

(NCMIP 2014)

Journal of Physics: Conference Series Volume 542

Cachan, France
23 May 2014

ISBN: 978-1-63439-466-6
ISSN: 1742-6588

Printed from e-media with permission by:

Curran Associates, Inc.
57 Morehouse Lane
Red Hook, NY 12571

Some format issues inherent in the e-media version may also appear in this print version.

Printed by Curran Associates, Inc. (2014)

Institute of Physics
Dirac House, Temple Back
Bristol BS1 6BE UK

Phone: 44 1 17 929 7481
Fax: 44 1 17 920 0979

techtracking@iop.org

Additional copies of this publication are available from:

Curran Associates, Inc.
57 Morehouse Lane
Red Hook, NY 12571 USA
Phone: 845-758-0400
Fax: 845-758-2634
Email: curran@proceedings.com
Web: www.proceedings.com

TABLE OF CONTENTS

STOCHASTIC DIFFUSION EQUATION WITH SINGULAR DIFFUSIVITY AND GRADIENT-DEPENDENT NOISE IN BINARY TOMOGRAPHY 1
B Sixou, L Wang, F Peyrin

INVERSION WITHOUT EXPLICIT JACOBIAN CALCULATIONS IN ELECTRICAL IMPEDANCE TOMOGRAPHY 7
A Fouchard, S Bonnet, L Hervé, O David

THREE-DIMENSIONAL TRANSIENT ELASTODYNAMIC INVERSION USING THE MODIFIED ERROR IN CONSTITUTIVE RELATION 13
M Bonnet, W Aquino

GOAL-ORIENTED UPDATING OF MECHANICAL MODELS 19
L Chamoin, P Ladevèze, J Waeytens

PHYSICAL MODELING OF MICROTUBULES NETWORK 25
P Allain, C Kervrann

AN ALTERNATING MINIMIZATION METHOD FOR BLIND DECONVOLUTION FROM POISSON DATA 31
M Prato, A La Camera, S Bonettini

PARTICLE MCMC FOR BAYESIAN MICROWAVE CONTROL 37
P Minvielle, A Todeschini, F Caron, P Del Moral

OPTICAL IMAGING IN A VARIATIONAL BAYESIAN FRAMEWORK 43
S Arhab, H Ayasso, B Duchêne, A Mohammad-Djafari

INFLUENCE OF PARTIALLY KNOWN PARAMETER ON FLAW CHARACTERIZATION IN EDDY CURRENT TESTING BY USING A RANDOM WALK MCMC METHOD BASED ON METAMODELING 49
C Cai, T Rodet, M Lambert

Author Index

Foreword

This volume of *Journal of Physics: Conference Series* is dedicated to the scientific contributions presented during the 4th International Workshop on New Computational Methods for Inverse Problems, NCMIP 2014 (`http://www.farman.ens cachan.fr/NCMIP_2014.html`). This workshop took place at Ecole Normale Supérieure de Cachan, on May 23, 2014. The prior editions of NCMIP also took place in Cachan, France, firstly within the scope of ValueTools Conference, in May 2011 (`http://www.ncmip.org/2011/`), and secondly at the initiative of Institut Farman, in May 2012 and May 2013, (`http://www.farman.ens-cachan.fr/NCMIP_2012.html`), `http://www.farman.ens-cachan.fr/NCMIP_2013.html`).

The New Computational Methods for Inverse Problems (NCMIP) Workshop focused on recent advances in the resolution of inverse problems. Indeed, inverse problems appear in numerous scientific areas such as geophysics, biological and medical imaging, material and structure characterization, electrical, mechanical and civil engineering, and finances. The resolution of inverse problems consists of estimating the parameters of the observed system or structure from data collected by an instrumental sensing or imaging device. Its success firstly requires the collection of relevant observation data. It also requires accurate models describing the physical interactions between the instrumental device and the observed system, as well as the intrinsic properties of the solution itself. Finally, it requires the design of robust, accurate and efficient inversion algorithms. Advanced sensor arrays and imaging devices provide high rate and high volume data; in this context, the efficient resolution of the inverse problem requires the joint development of new models and inversion methods, taking computational and implementation aspects into account. During this one-day workshop, researchers had the opportunity to bring to light and share new techniques and results in the field of inverse problems.

The topics of the workshop were: algorithms and computational aspects of inversion, Bayesian estimation, Kernel methods, learning methods, convex optimization, free discontinuity problems, metamodels, proper orthogonal decomposition, reduced models for the inversion, non-linear inverse scattering, image reconstruction and restoration, and applications (bio-medical imaging, non-destructive evaluation...).

NCMIP 2014 was a one-day workshop held in May 2014 which attracted around sixty attendees. Each of the submitted papers has been reviewed by two reviewers. There have been nine accepted papers. In addition, three international speakers were invited to present a longer talk.

The workshop was supported by Institut Farman (ENS Cachan, CNRS) and endorsed by the following French research networks (GDR ISIS, GDR MIA, GDR MOA, GDR Ondes). The program committee acknowledges the following research laboratories: CMLA, LMT, LURPA, SATIE.

Eric Vourc'h and Thomas Rodet

Journal of Physics: Conference Series **542** (2014) 011001 doi:10.1088/1742-6596/542/1/011001

Workshop co-chair:

Eric Vourc'h, SATIE laboratory, Ecole Normale Supérieure de Cachan, CNRS, France
Thomas Rodet, SATIE laboratory, Ecole Normale Supérieure de Cachan, CNRS, France

Technical program committee:
Alexandre Baussard, ENSTA Bretagne, Lab-STICC, France
Marc Bonnet, ENSTA, ParisTech, France
Laure Blanc-Féraud, I3S laboratory and INRIA Nice Sophia-Antipolis, France
Antonin Chambolle, CMAP, Ecole Polytechnique, CNRS, France
Oliver Dorn, School of Mathematics, University of Manchester, UK
Cécile Durieu, SATIE, ENS Cachan, CNRS, France
Gérard Favier, I3S Laboratory, University of Nice Sophia-Antipolis, France
Laurent Fribourg, LSV, ENS Cachan, CNRS, France
Jerôme Idier, IRCCyN Laboratory, Ecole Centrale de Nantes, France
Pierre-Yves Joubert, IEF, Paris-Sud University, CNRS, France
Marc Lambert, L2S Laboratory, CNRS, SupElec, Paris-Sud University, France
Dominique Lesselier, L2S Laboratory, CNRS, SupElec, Paris-Sud University, France
Matteo Pastorino, DIBE, University of Genoa, Italy
Christian Rey, LMT, ENS Cachan, CNRS, France

Local chair: Sophie Abriet, SATIE Laboratory, ENS Cachan, France
Veronique Almadovar, CMLA Laboratory, ENS Cachan, France
Béatrice Bacquet, SATIE Laboratory, ENS Cachan, France
Virginie Pauchont, CMLA Laboratory, ENS Cachan, France

Invited speakers
Andrea Massa, Universitá di Trento, ELEDIA Research Center, Trento, Italy.
Gabriel Peyre, Ceremade laboratory, University of Paris Dauphine, Paris, France.
Jean-Yves Tourneret, IRIT laboratory, ENSEEIHT, Toulouse, France.

Journal of Physics: Conference Series **542** (2014) 011002 doi:10.1088/1742-6596/542/1/011002

Peer review statement

All papers published in this volume of *Journal of Physics: Conference Series* have been peer reviewed through processes administered by the proceedings Editors. Reviews were conducted by expert referees to the professional and scientific standards expected of a proceedings journal published by IOP Publishing.

4th International Workshop on New Computational Methods for Inverse Problems (NCMIP2014) IOP Publishing
Journal of Physics: Conference Series **542** (2014) 012001 doi:10.1088/1742-6596/542/1/012001

Stochastic diffusion equation with singular diffusivity and gradient-dependent noise in binary tomography

B Sixou, L.Wang[1], F.Peyrin[2]
[1] CREATIS, INSA-Lyon; Inserm, U1044; CNRS UMR 5220; Université Lyon 1; Université de Lyon; 69621, Lyon, France
[2] ESRF, Imaging Group, 38043 Grenoble Cedex, France
E-mail: bruno.sixou@insa-lyon.fr

Abstract. In this work, we use stochastic diffusion equation with a singular diffusivity and a gradient-dependent noise to improve the reconstruction of binary tomography cross-sections obtained from a small number of projections. A first reconstruction image is obtained with the Total Variation regularization method. The reconstruction is then refined with this stochastic approach. The method is applied to a noisy bone cross-section with 10 projection angles.

1. Introduction

The tomographic reconstruction problem from few projections is an ill-posed problem which has many applications in medical imaging and material science. Binary tomography methods may be proposed to set a simpler inverse problem [1]. They are associated to an under-determined linear system of equations, $Rf = p^\delta$, with the Radon operator R, the measured projection values p^δ, and the pixel values $(f_i)_{1\leq i\leq n}$ of the image with binary constraints $f = (f_1, ...f_n) \in \{0,1\}^n$. Various approaches have been studied to solve this reconstruction problem [1, 2, 3] based on convex analysis optimization [4], Markov random fields [5] or Belief Propagation [6]. Very good reconstruction results are obtained with the Total Variation regularization [7]. Some errors are still present on the boundaries of the reconstructed regions. The main contribution of this work is to use stochastic nonlinear diffusion equation with singular diffusivity and a gradient dependent noise to improve the solution obtained with the TV regularization. After the introduction, the TV regularization method for binary tomography is presented. Then, we describe the nonlinear diffusion with a new existence result. The implementation of the method and the numerical results obtained on a noisy bone CT cross-section are reported in the last section.

2. Total Variation regularization and Alternate Direction of Minimization Method

The binary tomography problem is ill-posed and must be regularized. We consider a TV regularization[8, 9] with the convex constraints $f \in [0,1]^n$ and the following problem:

$$(P) \ \text{minimize} \ \frac{\mu}{2}\|p^\delta - Rf\|_2^2 + J_{TV}(f) \ \ s.t. \ \ f \ \in [0,1]^n \tag{1}$$

Better reconstruction results were obtained with the TV regularization than with the Tikhonov regularization. The parameter μ is the regularization parameter balancing the two terms and the noisy projections p^δ are such that $\|p^\delta - p\| \leq \delta$, where δ is the noise level and p

the projections without noise. For an image f $\in H_1(D)$, the regularization term is given by $J_{TV}(f) = \int_D |\nabla f(r)|dr$. A fast, efficient TV/L_2 minimization algorithm based on the Alternate Direction of Minimization Method (ADMM) has been proposed [8, 9]. In order to include convex constraints, $f \in C = [0,1]^n$, the following augmented Lagrangian $\mathcal{L}(f,(g_i),h,(\lambda_i),\lambda_C)$ is considered:

$$\mathcal{L} = \sum_i (\|g_i\|_2 - \lambda_i^t(g_i - D_i f) + \frac{\beta}{2}\|g_i - D_i f\|_2^2) + \frac{\mu}{2}\|g - Af\|_2^2 + I_C(h) + \frac{\beta}{2}\|h - f\|_2^2 - \lambda_C^t(h - f) \quad (2)$$

The Lagrange multipliers (λ_i), λ_c are vectors in $\mathbb{R}^{2n}$ and $\mathbb{R}^n$ and β is the Lagrangian parameter. For each pixel i, $D_i f \in \mathbb{R}^2$ represents the first-order finite difference in both horizontal and vertical directions. The sequences $(f^k,(g_i^k),h^k,(\lambda_i^k),\lambda_C^k)_{k\geq 0}$ generated by the ADMM algorithm converges to a Kuhn-Tucker point of problem (P),$(f^*,(g_i^*),h^*,(\lambda_i^*),\lambda_C^*)_{k\geq 0}$, if (P) has one. The new iterate f^{k+1} is obained from the linear system :

$$(\sum_i D_i^t D_i + \frac{\mu}{\beta}A^t A + I)f = \sum_i D_i^t(g_i^{k+1} - \frac{1}{\beta}\lambda_i^k\} + \frac{\mu}{\beta}A^t g + h^{k+1} - \frac{\lambda_C^k}{\beta} \quad (3)$$

3. Singular stochastic diffusion equation with gradient dependent noise

We describe the stochastic diffusion used to improve the TV reconstruction and show a new existence result. Let D is a bounded open subset of $\mathbb{R}^2$, $(V = H_0^1(D), \|.\|_1)$, and a filtered probability spaces $(\Omega, \mathcal{F}, \{\mathcal{F}\}_t, \mathcal{P})$. To refine the solution obtained with the ADMM algorithm, we consider the singular stochastic diffusion equation (E_1) on $(H = L_2(D), |.|)$ [10] of the type:

$$\left\{\begin{array}{lr} dX(t) = \text{div }(\text{sgn}(\nabla)(X(t))dt - R^*(RX(t) - p^\delta) + \sigma(\nabla X(t))dW(t) & in\ (0,\infty)\times D \\ X(t) = 0 & (0,T)\times\partial D \\ X(0) = x \text{ in } D & \end{array}\right. \quad (4)$$

The noise term σ is a gradient dependent noise $\sigma(\nabla X(t))dW(t) = \frac{\partial X}{\partial x}dW_1(x,y,t) + \frac{\partial X}{\partial y}dW_2(x,y,t)$ where $(W_k(x,y,t))_{k=1,2}$ are independent Wiener random fields with continuous covariance function $(r_k(x,y))_{k=1,2}$ bounded by a constant r_0. This type of equation has been extensively studied with additive and multiplicative noise [11]. The new equation with a gradient-dependent noise is considered in this work because the noise is located on the boundaries of the reconstructed images. The multi-valued maximal monotone function $u \to \text{sgn}(u)$ from $\mathbb{R}^d$ into $2^{\mathbb{R}^d}$ is defined by:

$$\text{sgn}(u) = \left\{\begin{array}{l} \frac{u}{|u|_d}\ if\ u \neq 0 \\ \{v \in \mathbb{R}^d\ |v| \leq 1\}\ if\ u = 0 \end{array}\right. \quad (5)$$

where $|u|_d$ is the Euclidean norm. Following Barbu et al. [11], to show some existence results, we consider the approximating equation (E_2), $dX(t) = A_\epsilon X(t) - R^*(RX - p^\delta) + \sigma(\nabla X(t))dW(t)$ where the operator $A_\epsilon : H \to H$ is defined by $A_\epsilon(u) = (1+\epsilon A)^{-1}div(\beta_\epsilon(\nabla(1+\epsilon A)^{-1})u)$ with $A = -\Delta$. The operator β_ϵ is the Yosida approximation of the sgn(u) function:

$$\beta_\epsilon(u) = \left\{\begin{array}{l} \frac{u}{\epsilon}\ if\ |u|_d \leq \epsilon \\ \frac{u}{|u|_d}\ if\ |u|_d > \epsilon \end{array}\right. \quad (6)$$

3.1. *Existence and unicity of a solution of equation E_2*

We first show that the approximating equation is well-defined and that it is possible to use the variational approach for monotone and coercive stochastic partial differential equations [10]. Let $B = A_\epsilon + R^*R$, this operator is linear and weakly continous. We first show that:

$$\left\{\begin{array}{c} \exists\ \omega > 0,\ \omega_1 \text{ such that } \forall u \in V \\ < Bu, u > + |\sigma(u)|_r^2 \leq \omega_1|u|^2 - \omega\|u\|_1^2\ a.e\ (t,\omega) \end{array}\right. \quad (7)$$

where $|.|_r$ is the norm induced by the covariance operator r. Recalling that $\beta_\epsilon(u).u \geq |u|_d^2$ for $u \in V$, it is possible to choose ϵ such that

$$< A_\epsilon u, u >= - < \beta_\epsilon(\nabla(1+\epsilon A)^{-1})u), \nabla(1+\epsilon A)^{-1})u >$$
$$\leq -|\nabla(1+\epsilon A)^{-1}u|^2 \leq \frac{-|\nabla u|^2}{2}$$

We can also obtain $< R^*Ru, u > \leq \|R\|^2|u|^2$ and

$$|\sigma(u)|_r^2 = \sum_{j=1}^{2}\int_D r_j(x,x)|\frac{\partial u}{\partial x_j}|^2 dx \leq r_0|\nabla u|^2 \leq r_0\|u\|_1^2 \quad (8)$$

The former coercivity condition holds provided r_0 is small enough. We now show the monotonicity condition is satisfied:

$$\begin{cases} \text{for some constant } \eta \;\; \forall u, v \in V \\ 2 < B(u-v), u-v > + |\sigma(u)-\sigma(v)|_r^2 \leq \eta|u-v|^2 \;\; a.e \end{cases} \quad (9)$$

Let u,v $\in$ V, since the projection operator R is injective we obtain

$$- < R^*R(u-v), u-v >= - < R(u-v), R(u-v) > \leq -\lambda_{min}|u-v)|^2 \quad (10)$$

where λ_{min} is the minimal singular value of R. Thus, for $r_0 < 1/2$

$$\begin{aligned} &2 < B(u-v), u-v > + \|\sigma(u)-\sigma(v)\|_R^2 \\ &\leq (r_0 - 1/2)\|\nabla(u-v)\|_1^2 - \lambda_{min}|u-v|^2 \leq -\lambda_{min}|u-v)|^2 \;\; a.e \;\; (t,\omega) \end{aligned} \quad (11)$$

The monotonicity condition is satisfied. By standards existence results [10], the equation (E_2) has a unique solution $X_\epsilon \in \mathcal{C}_W([0,T];H)$ which is path-wise H continuous, where $\mathcal{C}_W([0,T];H)$ is the space of continous functions from $[0,T]$ to $L^2(\Omega;H)$ which are adapted to $\{\mathcal{F}\}_t$.

3.2. Existence of a solution of equation (E_1)

In order to control the gradient noise term, we introduce a Banach space with the norm [10], $|u|_T = E\{\sup_{0\leq t\leq T}|u|^2 + \int_0^T \|u\|_1^2 dt\}$. Let $\tilde{X_\epsilon} = (1+\epsilon A)^{-1}X_\epsilon$, $\tilde{X_\lambda} = (1+\lambda A)^{-1}X_\lambda$, by Ito's formula [11], we have:

$$\begin{aligned} &\frac{1}{2}|X_\epsilon - X_\lambda|^2 + \int_0^t\int_\Omega < \beta_\epsilon(\nabla\tilde{X_\epsilon}) - \beta_\lambda(\nabla\tilde{X_\lambda}), \epsilon\beta_\epsilon(\nabla\tilde{X_\epsilon}) - \lambda\beta_\lambda(\nabla\tilde{X_\lambda}) > dxds \\ &\leq \int_0^t (\sigma(X_\epsilon) - \sigma(X_\lambda), X_\epsilon - (X_\lambda)dW(s) + \int_0^t |\sigma(X_\epsilon) - \sigma(X_\lambda|_r^2 ds \end{aligned} \quad (12)$$

With the Davis-Burkholder-Gundy inequality, we obtain that there is a constant c such that:

$$\begin{aligned} E\sup_{0\leq t\leq T}\int_0^t (\sigma(X_\epsilon) - \sigma(X_\lambda), X_\epsilon - (X_\lambda)dW(s) &\leq cE(\int_0^T (\sigma(X_\epsilon) - \sigma(X_\lambda), X_\epsilon - X_\lambda)_r^2 ds)^{1/2} \\ &\leq cE\sup_{0\leq t\leq T}|X_\epsilon - X_\lambda|\sqrt{\int_0^T |\sigma(X_\epsilon) - \sigma(X_\lambda)|_r^2 ds} \quad (13) \\ &\leq \frac{1}{4}E\sup_{0\leq t\leq T}|X_\epsilon - X_\lambda|^2 + c^2E\int_0^T |\sigma(X_\epsilon) - \sigma(X_\lambda)|_r^2 ds \end{aligned}$$

Since $|\beta_\epsilon|_\infty \leq 1$, we obtain

$$E \sup_{0\leq t\leq T} \frac{1}{4}|X_\epsilon - X_\lambda|^2 \leq \lambda + \epsilon + r_0(1+c^2)E\int_0^t \|X_\epsilon - X_\lambda\|_1^2 ds \leq \lambda + \epsilon + r_0(1+c^2)|X_\epsilon - X_\lambda|_t^2 \tag{14}$$

By integrating from 0 to T, there exist constants C_1, C_2, C_3, C_4 such that

$$E\int_0^T\int_0^t |\sigma(X_\epsilon) - \sigma(X_\lambda|_R^2 dsdt \leq C_1 E\int_0^T\int_0^t \|X_\epsilon - X_\lambda\|_1^2 dsdt \leq C_2 E\int_0^T |X_\epsilon - X_\lambda|_t^2 dt \tag{15}$$

$$\begin{aligned} &E\int_0^T\int_0^t (\sigma(X_\epsilon) - \sigma(X_\lambda), X_\epsilon - X_\lambda)dW(s)dv \\ &\leq C_3(E\int_0^T\int_0^t (\sigma(X_\epsilon) - \sigma(X_\lambda)^2 dW(s)dt + E\int_0^T\int_0^t (X_\epsilon - X_\lambda)^2 dW(s)dt \\ &\leq C_4\int_0^T |X_\epsilon - X_\lambda|_t^2 dt \end{aligned} \tag{16}$$

combining (15) and (16), we obtain: $\int_0^T |X_\epsilon - X_\lambda|_1^2 du \leq \lambda + \epsilon + C_2\int_0^T |X_\epsilon - X_\lambda|_t^2 dt$. Combining (14) and the last inequality, with the Gronwall lemma, we deduce that $|X_\epsilon - X_\lambda|_T^2 \leq C_5(\epsilon + \lambda)\ \ \forall\ \lambda, \epsilon > 0\ \ a.e$ and there is a continuous H-valued process $X \in \mathcal{C}_W([0,T];H)$ such that $\lim_{\epsilon\to 0}|X_\epsilon - X|_T = 0\ \ a.e.$

4. Results and discussion

4.1. Simulations details

The TV regularization and the stochastic method were tested with 10 projections angles on a 256x256 bone cross-section acquired with synchrotron micro-CT [12]. Figure 1.a displays the reconstruction obtained from 400 projections with 400 rays per projections with Filtered Back Projection. The discrete approximation of the Radon transform is the one implemented in the Matlab Toolbox. The projections were corrupted with a Gaussian noise with a PSNR of 5.7 dB. The reconstructed image f_0 obtained with TV after binarization and the corresponding error map are displayed in Figure 1.b and 1.c respectively. The regularization parameters were chosen with the Morozov discrepancy principle [13] applied to the grey level image. The stochastic optimization is then applied with intermittent diffusions separated by TV/ADMM minimizations. The ADMM runs are stopped when the regularization functional stagnates. At the end of each step, the image is binarized with a threshold 0.5. Numerically, equation (E_1) is implemented with a stochastic noise term added at each iteration to the iterate f_k given by the ADMM method (Eq.3). The discretization of the Brownian motions was performed with the classical Euler-Maruyama method[14]. For a time step δt, the noise term added to the iterate f_k given by the ADMM method is given by $\mu Z_k\sqrt{\delta t}$ where the $(Z_k)_{k\geq 0}$ are i.i.d Gaussian random variables in $\mathbb{R}^n$, $Z_k \sim \mathcal{N}(0, r)$. The parameter μ controls the noise strength and r is a covariance function with Fourier transform $\tilde{r}(k) = (|k|^2 + 1)^{-2}$.

4.2. Numerical results

The decrease of the data term term related to the binary image $\|Rf^k_{binary} - p^\delta\|$ is displayed on Fig.2, where f^k_{binary} is the binarization of the grey-level image. In order to have more quantitative results, the evolution of the missclassification rate is displayed on Figure 3 for the iterated TV regularization or the intermittent diffusion. A decrease of the misclassification rate as a function of the number of iterations is also obtained. The stochastic approach is more efficient that the iterated TV regularization. The image obtained with the stochastic diffusion with the lowest value of $\|Rf^k_{binary} - p^\delta\|$ and the corresponding error map are shown in Fig.4a and Fig.4b. The reconstruction errors on the boundaries of the homogeneous regions are reduced.

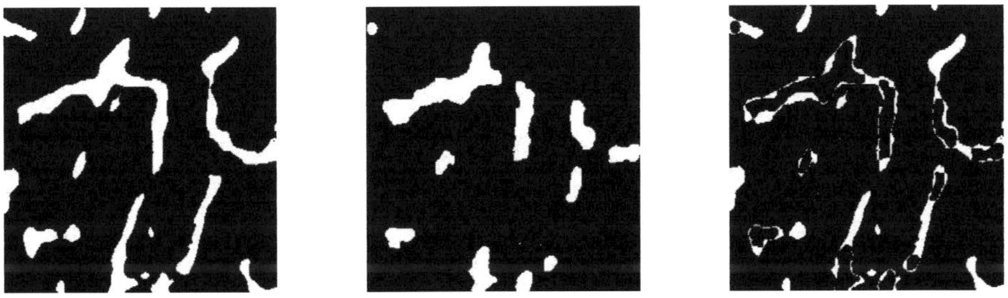

Figure 1. Reconstruction of the bone cross-section from 10 noisy projections with the TV regularization: (a) Ground truth;(b) Reconstructed image f_0; (c) Error map.

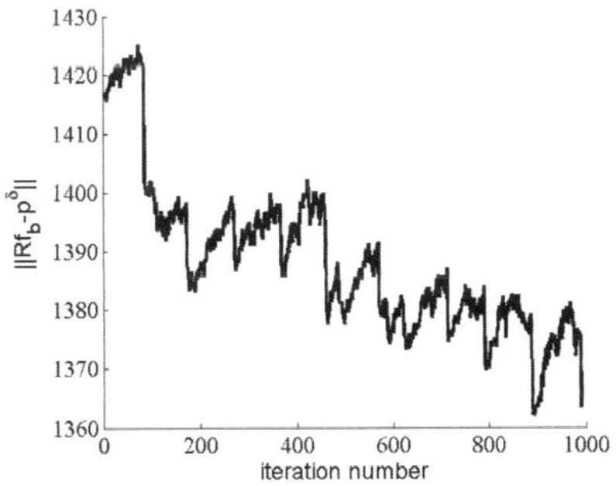

Figure 2. Evolution of the missclassification rate (i)iterated TV (iii) iterated stochastic diffusion.

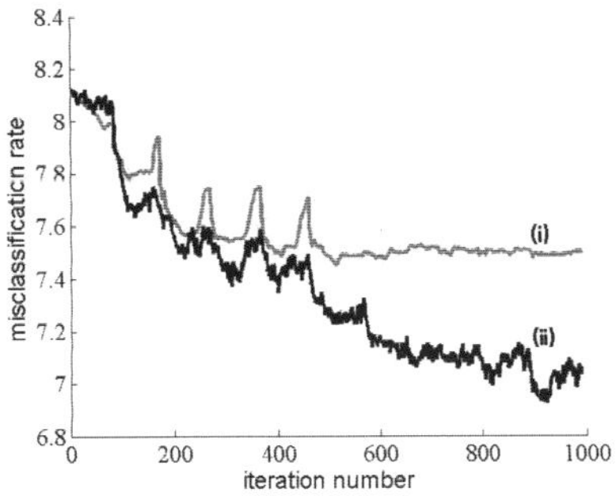

Figure 3. Evolution of the missclassification rate (i)iterated TV (iii) iterated stochastic diffusion.

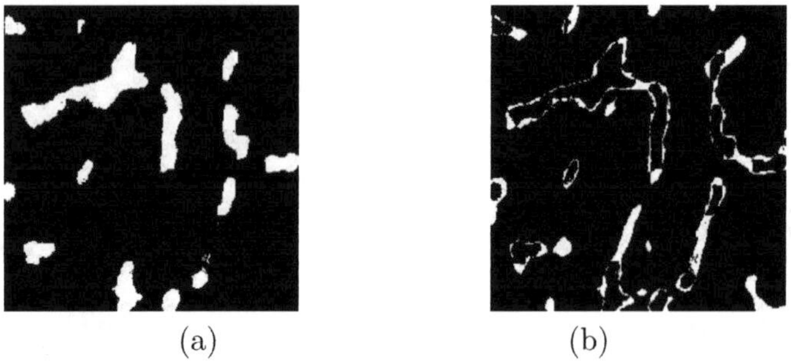

Figure 4. Reconstruction of the bone cross-section from 10 noisy projections with the iterated stochastic diffusion: (a) Reconstructed image; (b) Error map.

5. Conclusion

This work introduced a new stochastic method in tomographic reconstruction of for binary tomography from a limited number of views. The first reconstruction image is obtained with the Total Variation regularization. The reconstruction images are then refined with a stochastic non linear diffusion equation with a singular diffusivity and a gradient dependent noise. New existence results are demonstrated. The method is applied to a bone cross-section. The stochastic algorithm leads to a decrease of the reconstruction errors localized on the boundaries. The efficiency of the method will be investigated more precisely in future works.

References

[1] G.T.Herman and A.Kuba, "Advances in discrete tomography and its applications", Birkhauser Boston (2007).

[2] K.J.Batenburg and J.Sijbers, "Generic iterative subset algorithm for discrete tomography", *Discrete Applied Mathematics*, vol.157, pp. 438-451, 2009.

[3] W.Cai and L.Ma, "Comparaison of approaches based on optimization and algebraic iteration for binary tomography", *Computer Physics Communications*, vol.181, pp. 1974-1981, 2010.

[4] T.D.Capricelli and P.L.Combettes, "A convex programing algorithm for noisy discrete tomography", *Advances in discrete tomography and its applications*, Boston, MA, pp. 207-226, 2007.

[5] H.Y.Liao and G.T.Herman, "Automated estimation of the parameters of the Gibbs priors to be uses in binary tomography", *Discrete Applied Mathematics*, pp. 249-170, 2004.

[6] E.Gouillart, F.Krzakala, M.Mezard and L.Zdebrova, "Belief propagation reconstruction for discrete tomography", *Inverse Problems*, vol.29, pp. 035003, 2013.

[7] B.Sixou, L.Wang and F.Peyrin, *Binary tomographic reconstruction of bone microstructure from few projections with level-set regularization*, IEEE Symposium on Biomedical Imaging, San Francisco (2013).

[8] Y.Wang, J.Yang, W.Yin and Y.Zhang, "A new alternating minimization algorithm for Total Variation image reconstruction", *SIAM J.Imaging Sciences*, vol.1, pp. 248-272, 2008.

[9] M.Afonso, J.Bioucas-Dias and M.Figueiredo, "An Augmented Lagrangian approach to the constrained optimization formulation of imaging inverse problems", *IEEE Transactions on Image Processing*, vol.99, pp. 1: , 2009.

[10] C.Prevot and M.Rockner, "A concise course on stochastic partial differential equaions ", *Lecture Notes in Mathematics 1905* , 2007.

[11] V.Barbu, G.Da Prato, M.Rckner, "Stochastic nonlinear diffusion equations with singular diffusivity", *SIAM J.Math.Anal.*, vol.41, pp. 1106-1120, 2009.

[12] L.Apostol, V.Boudousq, O.Basset, C.Odet, S.Yot, J.Tabary,J.M.Dinten, E.Boller, P.O.Kotzki and F.Peyrin, "Relevance of 2D radiographic texture analysis for the assessment of 3D bone microarchitecture", *Medical Physics*, vol.33, pp. 3546-3556, 2006.

[13] H.W.Engl, M.Hanke and A.Neubauer, "Regularization of Inverse Problems", Dordrecht: Kluwer Academic, 1996.

[14] P.E.Kloeden and E.Platen, "Numerical solution of stochastic differential equations ", *Applications of Mathematics.* , vol.23, Berlin:Springer, 1992.

Inversion without Explicit Jacobian Calculations in Electrical Impedance Tomography

A Fouchard[1,2], S. Bonnet[1], L. Hervé[1], O. David[2]
[1]CEA, Leti, MINATEC Campus, 17 rue des Martyrs, F38054 Grenoble, Cedex 9, France
[2]Université Joseph Fourier, Grenoble Institute of Neuroscience, Bâtiment E.J. Safra, Chemin Fortuné Ferrini, 38700 La Tronche, France

E-mail: alexandre.fouchard@centraliens-lyon.net

Abstract. Electrical impedance tomography (EIT) is the inverse problem of finding the internal conductivity distribution of a medium given boundary electrical measurements performed via an electrode array onto its surface. Conventional inversion schemes adopt Tikhonov regularized Newton-type methods. Following a transport back-transport approach, we develop in this work an adjoint approach which allows reducing computational burden in enabling inversion without explicit Jacobian calculation. Forward and back-projection operators are defined from potential gradients, along with their efficient implementation. These derivations allow the transparent use of inversion algorithms. We first check the implementation of operators. We then evaluate how reconstructions perform on simulated noisy data using a preconditioned conjugate gradient. We eventually practice our inversion framework on experimental data acquired *in vitro* from a saline phantom.

1. Introduction

Biological tissues and fluids exhibit specific electrical properties, conductivity σ and permittivity ε, which may be used to infer their internal structure. Medical Electrical Impedance Tomography (EIT) is a soft-field, non-invasive imaging modality, which aims at reconstructing the distribution of internal electrical properties from boundary electrical measurements [1]. It works by injecting known low intensity alternating electrical currents through the body via an array of surface electrodes. The distribution of admittivity $\gamma = \sigma + j\omega\varepsilon$, at frequency ω, dictates internal current flows. Resulting voltages are measured between several pairs of electrodes.

EIT uses low frequency electrical currents from 1Hz to 10MHz. In this frequency range, current flows cannot be confined to a plane by a set of external electrodes; all measurements may be affected by a change in conductivity anywhere in the domain, not only those in the ray path. Hence, electrical imaging requires three dimensional modelling and needs adequate computational processing [2].

Using electrical measurements, reconstruction of admittivity distribution involves a two-step mathematical framework. The forward problem consists in building a numerical model able to predict expected measurements at the electrodes, given a distribution of admittivity. The inverse problem then estimates internal electrical properties through an optimization function which adjusts admittivity distribution in order to minimise the distance between measurements and calculated values.

Because the EIT inverse problem is non-linear and ill-posed, it cannot be easily solved. The standard reconstruction approach uses a family of regularized Newton-type methods [2] where admittivity distribution is represented using a piecewise constant model over voxels. Forward problem is linearized at some initial admittivity estimate and a Jacobian matrix is constructed. This matrix, also called sensitivity matrix, relates the admittivity of each internal voxel to the boundary measurements. Then, admittivity reconstruction involves the inversion of the ill-conditioned Jacobian matrix through generalized Tikhonov regularization. This requires inverting large linear systems of size equal to the number of voxels in the mesh times the number of measurements.

In order to optimise computational efficiency by avoiding explicit calculations of Jacobian, we propose here a transport back-transport method [3] that defines an adjoint reconstruction framework for EIT. Matrices involved in transport and back-transport operations present a reduced size by a factor of the number of electrodes, typically 16 or more in modern EIT systems. This approach opens new opportunities to use a wide range of inversion algorithms for EIT. It also offers a way to reduce complexity of EIT calculations.

In this paper, we focus only on conductivity reconstructions. The derivation of the forward and back-projection operators is presented. The reconstruction framework is then evaluated using the case of a 2D preconditioned conjugate gradient (PCG) inversion scheme. Conductivity reconstructions are performed against experimental data acquired *in vitro* on a saline phantom.

2. Methods

2.1. Sensitivity calculations

Optimization based methods are used in EIT to infer internal electrical property distribution. They require determining the derivative of the voltage measurements with respect to a conductivity parameter. The Jacobian matrix, or sensitivity matrix, contains all these partial derivatives.

The classic estimation of the Jacobian is based on the perturbation approach where two configurations are considered (Figure 1). The first configuration indexed by s is the actual measurement configuration, in which the source s is used. The second configuration indexed by d is a virtual measurement configuration in which source and detector have been interchanged.

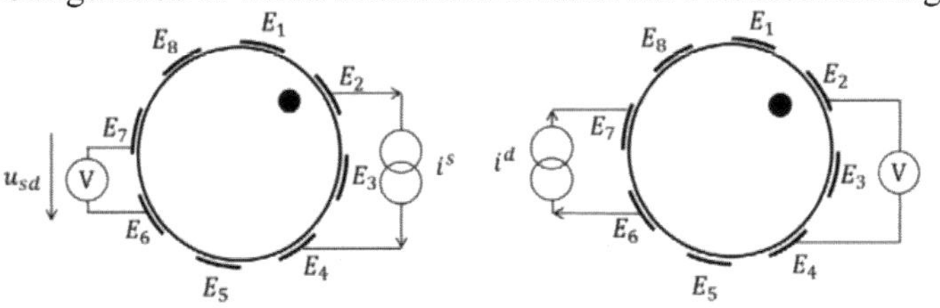

Figure 1. Actual (left) and virtual (right) measurement configurations for source and detector

Suppose the domain under study has been meshed into a partition of N_e elements. The perturbation of the voltage u_{sd}, measured in the situation where the source is s and detector d, due to a localized perturbation of conductivity in the element Ω_k, for a fixed injected current of amplitude i_{pp} is [4]:

$$\mathbf{S}_{k,sd} = \frac{\partial u_{sd}}{\partial \sigma_k} = \frac{-1}{i_{pp}} \int_{\Omega_k} \nabla v(i^s) \cdot \nabla v(i^d) \, \mathrm{d}\Omega \tag{1}$$

in which $\nabla v(i^s), \nabla v(i^d)$ are the potential gradients, respectively in actual and virtual configurations. With E electrodes, when rotating detector, we usually count E measurements per injection configuration: considering all measurements, the Jacobian matrix $\mathbf{J} = \mathbf{S}^T$ has dimension $E^2 \times N_e$.

Jacobian coefficients are calculated considering a piecewise linear potential, which leads to constant potential gradients per element. Hence, equation (1) reduces to:

$$\mathbf{S}_{k,sd} = \frac{-|\Omega_k|}{i_{pp}} \nabla v_k(i^s) \cdot \nabla v_k(i^d) \tag{2}$$

where $|\Omega_k|$ is the volume of element Ω_k in 3D. Potential gradients over each element of the mesh ∇v_k can be computed from standard finite element derivation of EIT physics [5].

2.2. Adjoint reconstruction framework

In this work, we follow a transport back-transport approach [3] which uses adjoint fields for reconstruction. Starting from an initial guess of conductivity, the framework calculates the difference between the computed and the actual measurements by solving a direct transport problem, and then transports these residuals back into the medium by solving a corresponding adjoint transport problem.

In order to avoid explicit calculations of the Jacobian and its adjoint, the Jacobian matrix $\mathbf{J}$ may be factorized under the form of equation (3) with matrices $\mathbf{G}_{k,s}^{x_i}, \mathbf{G}_{k,d}^{x_i} \in \mathbb{R}^{N_e \times E}$, given in equation (4), containing potential gradients relative to coordinate x_i, respectively for source injection configurations and virtual detector injection configurations.

$$\mathbf{J}_{sd,k} = -\sum_{x_i} \mathbf{G}_{k,s}^{x_i} \mathbf{G}_{k,d}^{x_i} \tag{3}$$

$$\mathbf{G}_{k,s}^{x_i} = \mathrm{diag}\left(\sqrt{\frac{|\Omega|}{i_{pp}}}\right) \begin{bmatrix} \nabla_{x_i} v_1(i^1) & \cdots & \nabla_{x_i} v_1(i^E) \\ \vdots & \ddots & \vdots \\ \nabla_{x_i} v_{N_e}(i^1) & \cdots & \nabla_{x_i} v_{N_e}(i^E) \end{bmatrix} \tag{4}$$

This factorization leads to express the direct and adjoint transport operators without explicit calculations of Jacobians but only of matrices $\mathbf{G}_{k,s}^{x_i}$ and $\mathbf{G}_{k,d}^{x_i}$, possibly equal.

2.2.1. Direct transport: forward operator The direct transport problem predicts expected measurements given a map of conductivity. It performs the operation $\delta \mathbf{u} = \mathbf{J} \delta \boldsymbol{\sigma}$. From the perturbation formula of equation (2) under the assumption of piecewise linear potential, the operation can be written in the formalism which uses gradient matrices, with $\delta \mathbf{U} \in \mathbb{R}^{E \times E}$:

$$\delta \mathbf{U}_{s,d} = -\sum_{x_i} \sum_k \mathbf{G}_{k,s}^{x_i} \mathbf{G}_{k,d}^{x_i} \delta \boldsymbol{\sigma}_k \tag{5}$$

2.2.2. Adjoint transport: back-projection operator The adjoint transport problem transports back to the image $\delta \boldsymbol{\sigma}$ the residuals $\mathbf{R} \in \mathbb{R}^{E \times E}$ between computed and measured values. It performs the adjoint operation $\delta \boldsymbol{\sigma} = \mathbf{J}^T vec(\mathbf{R})$. Using gradient matrices, it follows:

$$\delta \boldsymbol{\sigma}_k = -\sum_{x_i} \sum_s \sum_d \mathbf{G}_{k,s}^{x_i} \mathbf{G}_{k,d}^{x_i} \mathbf{R}_{s,d} \tag{6}$$

2.3. Implementation of inversion

The adjoint reconstruction framework depicted in the previous section allows using numerous inversion algorithms. In this work, standard PCG algorithm is used. CG features intrinsic regularizing properties and offers a good compromise between robustness, convergence, computation and image quality. Left preconditioning is used to weight measurements by the noise covariance matrix.

Numerical simulations are conducted using standard Finite Element Method in EIT with the Complete Electrode Model [1]. Custom Matlab routines adapted from the EIDORS library [6] allow the calculations of potential gradients along each mesh element and the implementations of forward and back-projection operators.

2.4. Experimental device

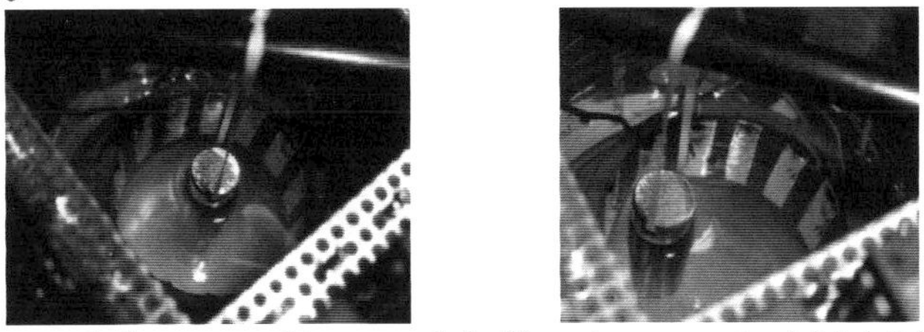

Figure 2. Experimental phantom with centred (left) and non-centred (right) inhomogeneity

Experimental measurements are performed on a saline phantom of 4cm diameter featuring 14 equally-spaced copper electrodes (Figure 2) with a custom-built EIT system [7]. Measurements are carried out using a four-electrode method with an injected current of 100μApp at 98kHz. Inhomogeneities are created by the presence of metallic cylinders of either 1cm or 5mm diameter, i.e. $1/4$ or $1/8$ of phantom diameter. A 5mm diameter hollow insulating cylinder is also used.

2.5. Reconstruction approach
Numerical derivations depicted in previous sections are first validated on simple test cases in 2D before considering image reconstruction of simulated and experimental measurements. A 2D mesh consisting of $N_e = 6{,}713$ elements will be used throughout the rest of this paper.

2.5.1. Implementation validation Two reference configurations, similar to the ones in Figure 2, are used to validate both forward and back-projection implementations. Background conductivity is 1S/m. Inclusions either centred or decentred exhibit a conductivity contrast of 0.1. Using the EIDORS library [6], the Jacobian $\mathbf{J}$ is calculated along with the residuals, obtained by inversion of the FEM admittance matrix. Proposed operators can be checked by comparing their results with the application of $\mathbf{J}$ or $\mathbf{J}^T$.

2.5.2. Simulated noisy measurements Difference reconstructions are conducted on simulated numerical measurements, perturbed with a 5% additive Gaussian noise on every measurement. These reconstructions allow studying the reconstruction framework behavior in presence of noise. The mesh used for the simulation of measurements (18,647 elements) is different from the mesh used for reconstructions (6,713 elements), in order to avoid committing an inverse crime [2].

2.5.3. Experimental measurements Difference reconstructions are performed given sets of data acquired from the experimental device. Data collection is performed 10 times in order to evaluate the mean value of measurements and their variance. The latter is incorporated in the left preconditioner.

3. Results

3.1. Validation of reconstruction framework implementation
Comparison between calculated gradients and theoretical ones, for a given potential distribution $v(x_1, x_2) = e^{-(x_1^2+x_2^2)}$, shows a mean value of 1 and a standard deviation of 0.05. This indicates that the implementation of elemental gradient calculations works fine. Both projection and back-projection ratios exhibit no discrepancies (mean value of 1 and corresponding standard deviation inferior to 10^{-11}), which validates the right implementation of corresponding operators.

3.2. 2D simulation results
Difference reconstructions are performed against simulated measurements (Figure 3).

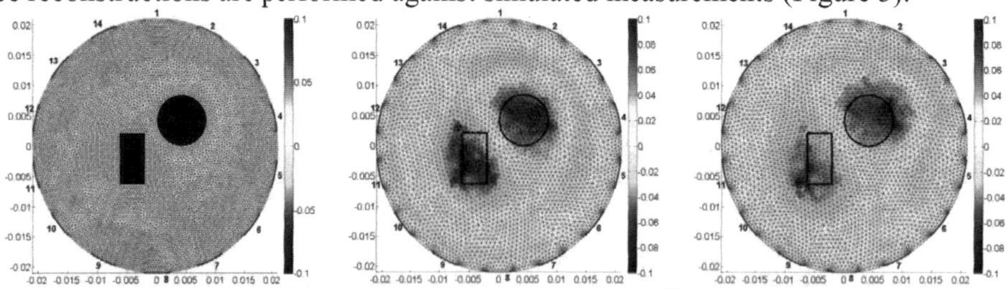

Figure 3. Difference reconstruction of simulated measurements;
left: conductivity to be imaged; centre: no noise; right: 5% additive Gaussian noise

Without noise, the PCG algorithm performs well to reconstruct both position and conductivity contrast of inclusions. With 14 electrodes, the geometrical details of the rectangular inclusion cannot be recovered, it is a classical behaviour of EIT [1]. In presence of noisy data, the PCG algorithm is able to detect the inclusions along with their conductivity contrast. But, spatial resolution has been lost in the reconstruction process, which is also a known feature of EIT [1]. Future developments of the adjoint framework presented in this work might consider inversion algorithms that favour sparse solutions.

The back-projection operator readily offers determination of sensitivity maps for any desired source-detector configuration (Figure 4).

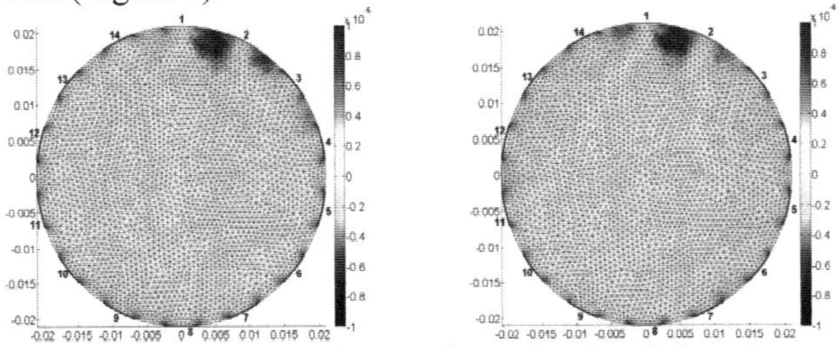

Figure 4. Sensitivity maps for the injection configuration involving electrodes 1 and 2; Measurement configuration involve electrodes 4 and 8 (left), electrodes 7 and 11 (right)

These maps present both positive sensitivity areas (yellow, red) and negative ones (blue). The latter explains the lack of spatial resolution put forward in previous reconstruction in the presence of noise.

3.3. Phantom experiments

Difference reconstructions from experimental data are now performed. Experiments are first led with conductive inclusion ($1/4$ of phantom diameter) featuring a contrast of 10^6 with the saline solution background (Figure 5).

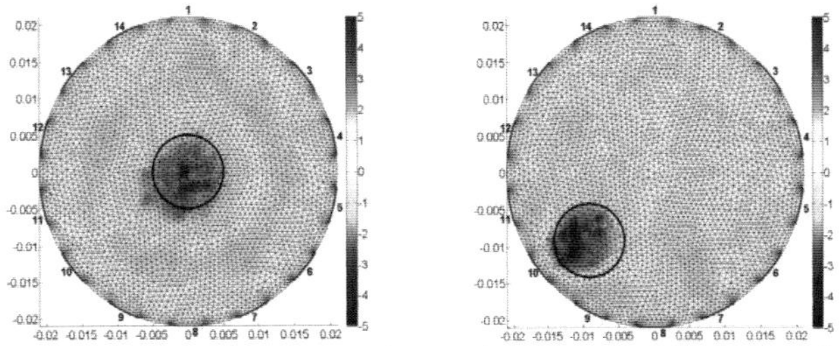

Figure 5. Difference reconstructions from experimental measurements on conductive inclusion; left: centred inclusion; right: non-centred inclusion; black circles: true position of inclusion

The PCG algorithm and subsequent forward and back-projection operators perform well in experimental conditions, with an estimated maximum noise of 5% in the corresponding measurements. Reconstruction drift in position seems to be handled quite properly.

Experiments are then conducted with two conductive inclusions ($1/4$ and $1/8$ of phantom diameter) or with a hollow insulating cylinder ($1/8$ of phantom diameter) (Figure 6).

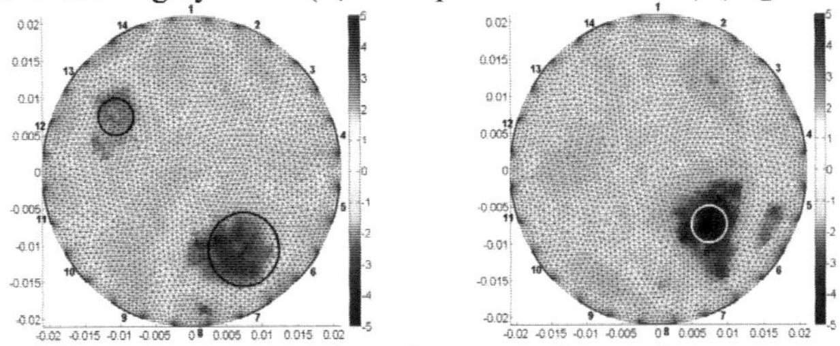

Figure 6. Difference reconstructions from experimental measurements; left: conductive inclusions; right: hollow insulating inclusion; circles: true position of inclusions

The framework behaves properly on both conductivity distributions. Recovering an inclusion whose dimensions equal $1/8$ of phantom diameter almost matches the spatial resolution limit of 14 electrode EIT.

Sharp contrasts are not recovered in this work since it requires using non-linear inversion algorithms.

4. Discussion

In this paper, we presented an inversion scheme which does not require explicit calculations of Jacobian. Instead, we practice a transport back-transport framework using an adjoint method and define two operators: forward and back-projection. Both can be efficiently determined from elemental gradient matrices detailed in previous sections. Such an approach allows using a wide range of inversion algorithms in the reconstruction process of EIT.

The validity of the implementation we propose was first checked. We then ran 2D difference reconstructions on simulated numerical data in which a 5% additive Gaussian noise was added. Classical behaviours of EIT reconstruction were observed, namely the loss of spatial resolution with noisy measurements. Eventually, we tested the method on experimental data acquired *in vitro* on a saline phantom.

In this scope, the framework allows seamlessly to determine estimated conductivity profiles and measurement configuration sensitivity maps. With the 14 electrode phantom used in this work, gains in matrices size used in calculations involve a reduction of 14 in one dimension, i.e. manipulated matrices were of size $6{,}713\times14$ instead of standard $6{,}713\times14^2$.

During the inversion process, the PCG algorithm easily allows incorporating prior information about the solution. In this work, measurement variance was incorporated in the left preconditioner. Consequent images prove to be qualitatively superior to the ones obtained using conventional regularization schemes, i.e. generalized Tikhonov. This aspect deserves more investigation.

Further studies will translate these developments in 3D and work on recovering more complicated conductivity distributions. Future developments of the adjoint framework presented in this work might consider reconstruction processes that favour sparse solutions and use non-linear inversion algorithms.

Acknowledgments

Authors would like to thank instrumentation engineer A. Noca from CEA-LETI for his noteworthy contribution on the design and setting-up of the experimental device.

References

[1] Holder D S (ed) 2004 *Electrical Impedance Tomography: Methods, History and Applications* (London: Taylor and Francis)

[2] Lionheart W R B 2004 EIT reconstruction algorithms: pitfalls, challenges and recent developments *Physiol. Meas.* **25** 125

[3] Dorn O 1998 A transport-backtransport method for optical tomography *Inverse Problems* **14**(5) 1107-30

[4] Geselowitz D B 1971 An Application of Electrocardiographic Lead Theory to Impedance Plethysmography *IEEE Trans. Biomed. Eng.* **18** 38–41

[5] Vavasis S A 1996 Stable finite elements for problems with wild coefficients *SIAM J. Numer. Anal.* **33** 890–916

[6] Adler A and Lionheart W R B 2006 Uses and abuses of EIDORS: an extensible software base for EIT *Physiol. Meas.* **27** S25

[7] Fouchard A, Noca A, Bonnet S, Pham P, Clarençon D and David O 2014 Modular architecture of a Multi-Frequency Electrical Impedance Tomography system: design and implementation submitted to *IEEE EMBC*

Three-dimensional transient elastodynamic inversion using the modified error in constitutive relation

Marc Bonnet[1], **Wilkins Aquino**[2]

[1] POems (UMR 7231 CNRS-INRIA-ENSTA), ENSTA, Palaiseau, France
[2] Department of Civil and Environmental Engineering, Duke University, Durham, USA

E-mail: `mbonnet@ensta.fr`, `wa20@duke.edu`

Abstract. This work is concerned with large-scale three-dimensional inversion under transient elastodynamic conditions by means of the modified error in constitutive relation (MECR), an energy-based cost functional. A peculiarity of time-domain MECR formulations is that each evaluation involves the computation of two elastodynamic states (one forward, one backward) which moreover are coupled. This coupling creates a major computational bottleneck, making MECR-based inversion difficult for spatially 2D or 3D configurations. To overcome this obstacle, we propose an approach whose main ingredients are (a) setting the entire computational procedure in a consistent time-discrete framework that incorporates the chosen time-stepping algorithm, and (b) using an iterative SOR-like method for the resulting stationarity equations. The resulting MECR-based inversion algorithm is demonstrated on a 3D transient elastodynamic example involving over 500,000 unknown elastic moduli.

Introduction. The identification of spatially-varying mechanical material parameters (e.g. elastic moduli) from measured elements of the mechanical response of the solid body is an important inverse problem arising in connection with e.g. seismic exploration, biomechanical imaging or damage detection. Considerable research has been dedicated to formulate algorithms for material identification, most often based on minimizing a L^2 cost functional [1, 5, 11].

Alternatively, energy-based cost functionals are quite relevant for this kind of inverse problem.They have been initially introduced in e.g. [8] for electrical impedance tomography and in [9] for error estimation in the linear elastic FEM. Such functionals, often referred in solid mechanics as error in constitutive relation (ECR) functionals, then quickly thereafter proved very useful for various mechanical parameter identification problems [4, 6, 7, 10]. In particular, modified ECR functionals (e.g. [10]), where reliable informations (equilibrium equations, initial conditions, known boundary conditions) are enforced strictly while unreliable informations (measured data, constitutive properties) are incorporated as constitutive or observation residuals, have become widely used for mechanical parameter identification problems.

In most available investigations using (M)ECR functionals, either (quasi-)static or frequency-domain conditions are assumed; moreover, the considered parameter identification problems are of moderate dimension. Recent efforts towards broadening that scope include large-scale inversion in a frequency-domain setting [3] and MECR functionals for parameter identification under transient dynamical conditions [6]. The latter approach entails, for each evaluation of the MECR functional, solving a set of stationarity equations governing one forward elastodynamic state (with initial conditions) and one backward elastodynamic state (with final conditions). Moreover, the two states are coupled, unlike in the case of L^2 minimization, which creates a major computational bottleneck [6], causing past investigations to be restricted to 1D configurations.

This work aims at formulating and demonstrating a MECR-based approach suitable for large-scale three-dimensional inversion under transient elastodynamic conditions. The forward-backward coupling bottleneck is overcome by (a) setting up and exploiting the stationarity conditions in a time-discrete framework that incorporates the chosen time-stepping algorithm, and (b) an iterative SOR-like treatment for the resulting stationarity equations.

Problem setting. Let a solid elastic body occupy the bounded and connected domain $\Omega \subset \mathbb{R}^d$ $(1 \leq d \leq 3)$ with boundary Γ. Transient motions of the solid during a time interval of interest $t \in [0, T]$ are governed by (a) the balance equations

$$\nabla \cdot \boldsymbol{\sigma} + \boldsymbol{f} = \rho \ddot{\boldsymbol{u}} \quad \text{in } \Omega \times [0, T], \qquad \boldsymbol{\sigma} \cdot \boldsymbol{n} = \boldsymbol{g} \qquad \text{on } \Gamma_{\mathrm{N}} \times [0, T], \tag{1}$$

where $\boldsymbol{u}$ is the displacement field, ρ is the mass density, $\boldsymbol{\sigma}$ denotes the stress tensor, $\boldsymbol{f}$ is a given body force density, $\boldsymbol{g}$ and $\Gamma_{\mathrm{N}} \subset \Gamma$ are the given surface force density (traction) and its support, and $\boldsymbol{n}$ is the unit normal vector pointing outward from Ω; (b) the kinematic compatibility equations

$$\boldsymbol{\varepsilon}[\boldsymbol{u}] = \tfrac{1}{2}(\nabla \boldsymbol{u} + \nabla \boldsymbol{u}^{\mathrm{T}}), \quad \text{in } \Omega \times [0, T], \qquad \boldsymbol{u} = \boldsymbol{0} \quad \text{on } \Gamma_{\mathrm{D}} \times [0, T], \tag{2}$$

where $\boldsymbol{\varepsilon}[\boldsymbol{u}]$ denotes the linearized strain tensor associated with the displacement $\boldsymbol{u}$ and $\Gamma_{\mathrm{D}} = \Gamma \backslash \Gamma_{\mathrm{N}}$ is the constrained part of Γ; (c) the homogeneous initial conditions

$$\boldsymbol{u}(\cdot, 0) = \dot{\boldsymbol{u}}(\cdot, 0) = \boldsymbol{0} \quad \text{in } \Omega; \tag{3}$$

and (d) the constitutive (linear elastic) relation, involving the fourth-order elasticity tensor $\boldsymbol{\mathcal{C}}$:

$$\boldsymbol{\sigma} = \boldsymbol{\mathcal{C}} : \boldsymbol{\varepsilon} \qquad \text{in } \Omega \times [0, T]. \tag{4}$$

Inverse problem. Supplementing the basic equations (1)–(4), additional information in the form of measured transient displacements $\widetilde{\boldsymbol{u}}$ in $\Omega_{\mathrm{m}} \times [0, T]$ (where $\Omega_{\mathrm{m}} \subset \Omega$ is a measurement region) is assumed to be available. The inverse problem then consists in finding the spatial distribution of the elastic moduli $\boldsymbol{\mathcal{C}}$ from the data $\widetilde{\boldsymbol{u}}$, subject to equations (1)–(4) with given excitations $\boldsymbol{f}, \boldsymbol{g}$.

Modified error in constitutive relation approach. Measurement noise is to be expected, in which case the exact enforcement of experimental data is normally not desirable. Accordingly, a modified error in constitutive relation (MECR) functional [2, 6] is defined by treating the discrepancy between measured quantities and their computed counterparts as a penalty term:

$$\widetilde{U}(\boldsymbol{u}, \boldsymbol{\sigma}, \boldsymbol{\mathcal{C}}) = \frac{1}{2} \int_0^T \big(\boldsymbol{\sigma}_t - \boldsymbol{\mathcal{C}} : \boldsymbol{\varepsilon}[\boldsymbol{u}_t] \,,\, \boldsymbol{\mathcal{C}}^{-1} : (\boldsymbol{\sigma}_t - \boldsymbol{\mathcal{C}} : \boldsymbol{\varepsilon}[\boldsymbol{u}_t]) \big) \, \mathrm{d}t + \frac{\xi f}{2} \| \boldsymbol{u} - \widetilde{\boldsymbol{u}} \|^2_{L^2(\Omega_{\mathrm{m}})} \tag{5}$$

where $(\boldsymbol{f}, \boldsymbol{g})$ stands for the $L^2(\Omega)$ scalar product of $\boldsymbol{f}$ and $\boldsymbol{g}$, $\xi > 0$ is a tunable dimensionless weight and f a fixed constant ensuring consistent units.

In this framework, the inverse problem is cast as the optimization problem

$$(\boldsymbol{u}^\star, \boldsymbol{\sigma}^\star, \boldsymbol{\mathcal{C}}^\star) = \arg \min_{\boldsymbol{u}, \boldsymbol{\sigma}, \boldsymbol{\mathcal{C}}} \widetilde{U}(\boldsymbol{u}, \boldsymbol{\sigma}, \boldsymbol{\mathcal{C}}) \qquad \text{subject to (1)–(3)}, \tag{6}$$

whose solution $(\boldsymbol{u}^\star, \boldsymbol{\sigma}^\star, \boldsymbol{\mathcal{C}}^\star)$ satisfies equations (1) to (3) (i.e. balance, compatibility and initial conditions) while achieving a compromise between (i) satisfying the constitutive relation (4) and (ii) matching the measurements $\widetilde{\boldsymbol{u}}$.

When using MECR functionals of the form (5), the mismatch between an assumed material property $\boldsymbol{\mathcal{C}}$ and measurements $\widetilde{\boldsymbol{u}}$ is quantified through the modified error in constitutive relation $\widetilde{E}(\boldsymbol{\mathcal{C}}) := \widetilde{U}(\boldsymbol{u}^{\boldsymbol{\mathcal{C}}}, \boldsymbol{\sigma}^{\boldsymbol{\mathcal{C}}}, \boldsymbol{\mathcal{C}})$, where the mechanical fields $(\boldsymbol{u}^{\boldsymbol{\mathcal{C}}}, \boldsymbol{\sigma}^{\boldsymbol{\mathcal{C}}})$ are determined by the partial minimization of $\widetilde{U}(\boldsymbol{u}, \boldsymbol{\sigma}, \boldsymbol{\mathcal{C}})$ for given $\boldsymbol{\mathcal{C}}$, and in particular depend on $\boldsymbol{\mathcal{C}}$:

$$(\boldsymbol{u}^{\boldsymbol{\mathcal{C}}}, \boldsymbol{\sigma}^{\boldsymbol{\mathcal{C}}}) := \arg \min_{\boldsymbol{u}, \, \boldsymbol{\sigma}} \widetilde{U}(\boldsymbol{u}, \boldsymbol{\sigma}, \boldsymbol{\mathcal{C}}) \qquad \text{subject to (1)–(3)}. \tag{7}$$

A natural approach for solving the minimization (6), previously used for frequency-domain problems in [3], then consists in an alternating-directions strategy whereby the transition from a current iterate $(\boldsymbol{u}, \boldsymbol{\sigma}, \boldsymbol{\mathcal{C}})$ to the next iterate $(\boldsymbol{u}^\star, \boldsymbol{\sigma}^\star, \boldsymbol{\mathcal{C}}^\star)$ consists in (i) setting $(\boldsymbol{u}^\star, \boldsymbol{\sigma}^\star) = (\boldsymbol{u}^{\mathcal{C}}, \boldsymbol{\sigma}^{\mathcal{C}})$, with $(\boldsymbol{u}^{\mathcal{C}}, \boldsymbol{\sigma}^{\mathcal{C}})$ solving the partial minimization (7), and (ii) finding the material update $\boldsymbol{\mathcal{C}}^\star$ by solving the partial minimization problem

$$\boldsymbol{\mathcal{C}}^\star := \arg\min_{\boldsymbol{\mathcal{C}}' \in \mathcal{C}} \widetilde{U}(\boldsymbol{u}^{\mathcal{C}}, \boldsymbol{\sigma}^{\mathcal{C}}, \boldsymbol{\mathcal{C}}'), \tag{8}$$

where $(\boldsymbol{u}^{\mathcal{C}}, \boldsymbol{\sigma}^{\mathcal{C}})$ are the outcome of (7) and are kept fixed. This approach has been implemented in this work and is employed for the numerical experiments.

Discrete formulation of MECR-based inversion. We now consider a time-discrete formulation of the transient dynamics involved in the definition and exploitation of the MECR functional, based on a Newmark time-stepping scheme. Introducing a sequence $t_0 = 0, t_1 = h, \ldots, t_k = kh, \ldots, t_n = Nh = T$ of discrete time instants (with a constant time step $h = T/N$ used for simplicity), we let f_k denote the value at $t = t_k$ of a generic time-dependent quantity f. The Newmark update relations linking the displacements $\boldsymbol{u}_k$, the velocities $\boldsymbol{v}_k := \dot{\boldsymbol{u}}_k$ and the accelerations $\boldsymbol{a}_k := \ddot{\boldsymbol{u}}_k$, which introduce two additional algorithmic parameters β, γ and read

$$\boldsymbol{u}_{k+1} = \boldsymbol{u}_k + h\boldsymbol{v}_k + h^2[\bar{\beta}\boldsymbol{a}_k + \beta\boldsymbol{a}_{k+1}], \qquad \boldsymbol{v}_{k+1} = \boldsymbol{v}_k + h[\bar{\gamma}\boldsymbol{a}_k + \gamma\boldsymbol{a}_{k+1}]. \tag{9}$$

(where $\bar{\beta} := \frac{1}{2} - \beta$ and $\bar{\gamma} := 1 - \gamma$, and with $\beta = 1/4$, $\gamma = 1/2$ used in the numerical experiments).

In this setting, introducing for convenience the shorthand notations $\boldsymbol{u}, \boldsymbol{v}, \boldsymbol{a}, \boldsymbol{\sigma}$ for denoting mechanical field histories $(\boldsymbol{u}_k, \boldsymbol{v}_k, \boldsymbol{a}_k, \boldsymbol{\sigma}_k)_{0 \le k \le N}$, the inverse problem consists in reconstructing the spatial distribution $\boldsymbol{\mathcal{C}}^\star$ of the elasticity tensor, and additionally in finding the mechanical field histories $\boldsymbol{\sigma}^\star$ and $(\boldsymbol{u}^\star, \boldsymbol{v}^\star, \boldsymbol{a}^\star)$, such that

$$(\boldsymbol{u}^\star, \boldsymbol{v}^\star, \boldsymbol{a}^\star, \boldsymbol{\sigma}^\star, \boldsymbol{\mathcal{C}}^\star) = \arg\min_{(\boldsymbol{u}, \boldsymbol{v}, \boldsymbol{a}) \in \mathbb{U},\, \boldsymbol{\sigma} \in \mathbb{S}(\boldsymbol{a}),\, \boldsymbol{\mathcal{C}}} \widetilde{U}_N(\boldsymbol{u}, \boldsymbol{\sigma}, \boldsymbol{\mathcal{C}}) \tag{10}$$

where the admissibility spaces $\mathbb{U}$ and $\mathbb{S}(\boldsymbol{a})$ respectively incorporate kinematic constraints (2), (3) and balance equations (1), and the time-discrete counterpart $\widetilde{U}_N$ of the MECR functional (5) is defined by

$$\widetilde{U}_N(\boldsymbol{u}, \boldsymbol{\sigma}, \boldsymbol{\mathcal{C}}) = \frac{1}{2}\sum_{k=0}^{N} \big(\boldsymbol{\sigma}_k - \boldsymbol{\mathcal{C}} : \boldsymbol{\varepsilon}[\boldsymbol{u}_k],\, \boldsymbol{\mathcal{C}}^{-1} : \boldsymbol{\sigma}_k - \boldsymbol{\varepsilon}[\boldsymbol{u}_k]\big) + \sum_{k=1}^{N} \frac{\xi f}{2} \|\boldsymbol{u} - \tilde{\boldsymbol{u}}\|^2_{L^2(\Omega_\text{m})}. \tag{11}$$

Lagrangian and optimality conditions. As problem (10) involves kinematical and dynamical admissibility constraints embedded in spaces $\mathbb{U}$ and $\mathbb{S}$, it is now reformulated in terms of its first-order Karush-Kuhn-Tucker necessary optimality conditions. The latter are derived from the following Lagrangian functional:

$$\begin{aligned}
\mathcal{L}(\boldsymbol{u}, \boldsymbol{v}, \boldsymbol{a}, \boldsymbol{\sigma}, \bar{\boldsymbol{u}}, \bar{\boldsymbol{v}}, \bar{\boldsymbol{a}}, \boldsymbol{\mathcal{C}}) &= \widetilde{U}_N(\boldsymbol{u}, \boldsymbol{\sigma}, \boldsymbol{\mathcal{C}}) + \big\langle \boldsymbol{a}_0, \rho\bar{\boldsymbol{u}}_0 \big\rangle + \big\langle \boldsymbol{\sigma}_0, \boldsymbol{\varepsilon}[\bar{\boldsymbol{u}}_0] \big\rangle - \mathcal{F}_0(\bar{\boldsymbol{u}}_0) - \big\langle \boldsymbol{\sigma}_0 \cdot \boldsymbol{n}, \bar{\boldsymbol{u}}_0 \big\rangle_{\Gamma \setminus \Gamma_\text{N}} \\
&+ \sum_{k=0}^{N-1} \Big\{ \big\langle \boldsymbol{\sigma}_{k+1}, \boldsymbol{\varepsilon}[\bar{\boldsymbol{u}}_{k+1}] \big\rangle + \big\langle \boldsymbol{a}_{k+1}, \rho\bar{\boldsymbol{u}}_{k+1} \big\rangle - \mathcal{F}_{k+1}(\bar{\boldsymbol{u}}_{k+1}) - \big\langle \boldsymbol{\sigma}_{k+1} \cdot \boldsymbol{n}, \bar{\boldsymbol{u}}_{k+1} \big\rangle_{\Gamma \setminus \Gamma_\text{N}} \\
&+ \big\langle \boldsymbol{u}_{k+1} - \boldsymbol{u}_k - h\boldsymbol{v}_k - h^2[\bar{\beta}\boldsymbol{a}_k + \beta\boldsymbol{a}_{k+1}], \rho\bar{\boldsymbol{a}}_{k+1} \big\rangle + \big\langle \boldsymbol{v}_{k+1} - \boldsymbol{v}_k - h[\bar{\gamma}\boldsymbol{a}_k + \gamma\boldsymbol{a}_{k+1}], \rho\bar{\boldsymbol{v}}_{k+1} \big\rangle \Big\}
\end{aligned} \tag{12}$$

which incorporates the MECR functional (11) together with the constraints resulting from (i) the initial balance equation, with multiplier field $\bar{\boldsymbol{u}}_0$; (ii) the current balance equations, with multiplier fields $\bar{\boldsymbol{u}}_{k+1}$ and (iii) the Newmark update equations (9), with multiplier fields $\bar{\boldsymbol{a}}_{k+1}$ and $\bar{\boldsymbol{v}}_{k+1}$. Moreover, the quantities $\bar{\boldsymbol{u}}, \bar{\boldsymbol{v}}, \bar{\boldsymbol{a}}$ are time-discrete sequences of Lagrange multiplier fields, $(\cdot, \cdot)$ denotes $L^2(\Omega)$ scalar products, and $\mathcal{F}_k$ are linear forms arising from the prescribed time-dependent excitations $\boldsymbol{f}, \boldsymbol{g}$ featured in (1).

The (first-order) necessary optimality conditions are found by setting to zero the first-order derivatives of the Lagrangian (12) with respect to (a) the stress history $\boldsymbol{\sigma}$, (b) the kinematic histories $(\boldsymbol{u}, \boldsymbol{v}, \boldsymbol{a}, \boldsymbol{\sigma})$, (c) the multiplier histories $(\bar{\boldsymbol{u}}, \bar{\boldsymbol{v}}, \bar{\boldsymbol{a}})$ and (d) the elasticity tensor $\boldsymbol{\mathcal{C}}$.

(a) Differentiating $\mathcal{L}$ with respect to $\boldsymbol{\sigma}$, the stress history in Ω is found to be given in terms of the kinematic histories by

$$\boldsymbol{\sigma}_0 = -\boldsymbol{\mathcal{C}}\!:\!\boldsymbol{\varepsilon}[\bar{\boldsymbol{u}}_0], \qquad \boldsymbol{\sigma}_k = \boldsymbol{\mathcal{C}}\!:\!\boldsymbol{\varepsilon}[\boldsymbol{u}_k - \bar{\boldsymbol{u}}_k] \quad (1 \le k \le N) \tag{13}$$

while the multiplier history is found to satisfy the boundary condition

$$\bar{\boldsymbol{u}}_k = \boldsymbol{0} \qquad \text{on } \Gamma_{\mathrm{D}} \qquad (0 \le k \le N), \tag{14}$$

(b) Setting to zero the derivatives of $\mathcal{L}$ with respect to the kinematic histories $(\boldsymbol{u}, \boldsymbol{v}, \boldsymbol{a})$, using (13) to eliminate stresses, and performing a finite element discretization, the following forward time-stepping equations are obtained for the space-discrete histories $(\boldsymbol{u}, \boldsymbol{v}, \boldsymbol{a})$ and $(\bar{\boldsymbol{u}}, \bar{\boldsymbol{v}}, \bar{\boldsymbol{a}})$:
   (i) Initial conditions:

$$\boldsymbol{u}_0 = \boldsymbol{v}_0 = \boldsymbol{0}\ ; \qquad \mathbf{M}\boldsymbol{a}_0 = \mathbf{K}\bar{\boldsymbol{u}}_0 + \mathbb{F}_0, \tag{15a}$$

   (ii) Forward time-stepping equations $(0 \le k \le N-1)$:

$$\mathbf{K}\boldsymbol{u}_{k+1} + \mathbf{M}\boldsymbol{a}_{k+1} = \mathbf{K}\bar{\boldsymbol{u}}_{k+1} + \mathbb{F}_{k+1}, \quad \text{and Newmark update equations (9).} \tag{15b}$$

(c) Setting to zero the derivatives of $\mathcal{L}$ with respect to the multiplier histories $(\bar{\boldsymbol{u}}, \bar{\boldsymbol{v}}, \bar{\boldsymbol{a}})$ simply restores all constraints introduced in $\mathcal{L}$. Upon elimination of stresses using (13) and finite element discretization, the following set of backward time-stepping equations is obtained.
   (i) Final conditions:

$$\mathbf{M}\bar{\boldsymbol{a}}_N\mathbf{K}\bar{\boldsymbol{u}}_N = -\xi f\mathbf{D}(\boldsymbol{u}_N - \widetilde{\boldsymbol{u}}_N), \qquad \bar{\boldsymbol{v}}_N = \boldsymbol{0}, \quad \bar{\boldsymbol{u}}_N - h^2\beta\bar{\boldsymbol{a}}_N = \boldsymbol{0}. \tag{16a}$$

   (ii) Backward transition equations $(1 \le k \le N-1)$:

$$\mathbf{M}\bar{\boldsymbol{a}}_k + \mathbf{K}\bar{\boldsymbol{u}}_k = \mathbf{M}\bar{\boldsymbol{a}}_{k+1} - \mathbf{K}\bar{\boldsymbol{u}}_{k+1} - \xi f\mathbf{D}(\boldsymbol{u}_k - \widetilde{\boldsymbol{u}}_k), \tag{16b}$$

$$\bar{\boldsymbol{v}}_k = \bar{\boldsymbol{v}}_{k+1} + h\bar{\boldsymbol{a}}_{k+1}, \qquad \bar{\boldsymbol{u}}_k - h\gamma\bar{\boldsymbol{v}}_k - h^2\beta\bar{\boldsymbol{a}}_k = h\bar{\gamma}\bar{\boldsymbol{v}}_{k+1} + h^2\bar{\beta}\bar{\boldsymbol{a}}_{k+1}. \tag{16c}$$

   (iii) Backward transition equation (last):

$$\bar{\boldsymbol{u}}_0 = h\bar{\gamma}\bar{\boldsymbol{v}}_1 + h^2\bar{\beta}\bar{\boldsymbol{a}}_1. \tag{16d}$$

Steps (a), (b) and (c) achieve the partial minimization (7), where $\boldsymbol{\mathcal{C}}$ is kept fixed. $\mathbf{K}$, $\mathbf{M}$, $\mathbb{F}_k$ respectively denote the stiffness and mass matrices arising from FE discretization, and the nodal excitation at time t_k, while the matrix $\mathbf{D}$ corresponds to L^2 scalar products on Ω_{m}.

(d) This step, constituting the second part of the alternating-direction approach, implements the partial minimization (8). Enforcing the first-order necessary optimality condition $\partial_{\boldsymbol{\mathcal{C}}} U(\boldsymbol{u}^{\boldsymbol{\mathcal{C}}}, \boldsymbol{\sigma}^{\boldsymbol{\mathcal{C}}}, \boldsymbol{\mathcal{C}})[\hat{\boldsymbol{\mathcal{C}}}] = 0\ \forall\hat{\boldsymbol{\mathcal{C}}}$ for the latter problem and assuming isotropic linear elasticity, the explicit pointwise updating formulas

$$B^\star = \left(\frac{\sum_{k=1}^N p_k^2}{\sum_{k=1}^N \epsilon_k^2} \right)^{1/2}, \qquad G^\star = \frac{1}{2}\left(\frac{\sum_{k=1}^N \boldsymbol{\sigma}_k^d : \boldsymbol{\sigma}_k^d}{\sum_{k=1}^N \boldsymbol{\varepsilon}^d[\boldsymbol{u}_k] : \boldsymbol{\varepsilon}^d[\boldsymbol{u}_k]} \right)^{1/2}. \tag{17}$$

are found, with B and G denoting (spatially-dependent) bulk and shear moduli, respectively. Since $B^\star, G^\star > 0$ by construction, any such update $\boldsymbol{\mathcal{C}}^\star$ of $\boldsymbol{\mathcal{C}}$ is positive definite.

Solution methodology for the stationarity equations. To evaluate $\widetilde{U}_N(\boldsymbol{\mathcal{C}})$ or perform the constitutive update (17) entails solving for the kinematic and multiplier histories the stationarity problem combining the forward equations (15a,b) and the backward equations (16a–d), which constitutes the cornerstone of MECR-based inversion. In addition to defining (as usual) evolution equation in opposite directions of time, the stationarity equations are here coupled. Practical considerations dictated by the large-scale identification objective suggest an iterative solution algorithm based on usual time-stepping solvers. Block successive over-relaxation (SOR) iterations are used here. Gathering equations (15a,b) and (16a–d) for all times steps, the forward and adjoint histories $X := (\boldsymbol{u}, \boldsymbol{v}, \boldsymbol{a})^{\mathrm{T}}$ and $\bar{X} := (\bar{\boldsymbol{u}}, \bar{\boldsymbol{v}}, \bar{\boldsymbol{a}})^{\mathrm{T}}$ can be considered as solving the following system of equations, set in block form:

$$\mathcal{A}X + \mathcal{C}\bar{X} = F, \qquad -\mathcal{D}X + \mathcal{A}^{\mathrm{T}}\bar{X} = G, \tag{18}$$

where the definition of matrices $\mathcal{A}, \mathcal{C}, \mathcal{D}$ and vectors F, G can be easily obtained by identification from (15a,b) and (16a–d). In particular, the matrix $\mathcal{A}$ is associated to the forward Newmark scheme, and the backward time-stepping can be shown to be represented by the adjoint matrix $\mathcal{A}^{\mathrm{T}}$. This implies in particular that the forward and backward time stepping schemes have identical stability conditions. Then, the iterative scheme is defined by introducing auxiliary unknowns $W, \bar{W}$ through

$$X^{(i+1)} = W + \bar{\eta} X^{(i)}, \qquad \bar{X}^{(i+1)} = \bar{W} + \bar{\eta} \bar{X}^{(i)}, \tag{19}$$

so that one iteration of the SOR algorithm consists, for given $(X^{(i)}, \bar{X}^{(i)})$ in solving equations

$$\text{(a)} \;\; \mathcal{A} W = \eta(F - \mathcal{C} X^{(i)}) \quad \text{then} \quad \text{(b)} \;\; \mathcal{A}^{\mathrm{T}} \bar{W} = \eta(G + \mathcal{D} X^{(i+1)}) \tag{20}$$

(where $0 < \eta < 2$ is the relaxation parameter) and obtain $(X^{(i+1)}, \bar{X}^{(i+1)})$ from (19) with the converged values of $W, \bar{W}$. The SOR iterations (20) are quite simple to implement using existing FE dynamical analysis codes, as they involve the usual stiffness and mass structural matrices. We used the finite element code SIERRA/SDA [13], which has massively parallel capabilities including efficient multilevel domain decomposition strategies for solving linear systems.

3D numerical example. A unit cube containing a cylindrical inclusion with an ellipsoidal cross section (Figure 1), its bottom surface being fixed, is loaded on its other faces with a normal uniform time-harmonic force density, whose magnitude is one unit on the top surface and 0.5 units on the side surfaces (load frequency 1 Hz, total load duration 1 s, with $N = 100$ time steps of $h = 0.01$ s). The true shear and bulk moduli of the background material are taken as $G = 2$ and $B = 4$, while those of the inclusion are $G = 6$ and $B = 16$. The 3D domain was discretized using $414,000$ tetrahedral finite elements ($75,000$ nodes). The data $\tilde{\boldsymbol{u}}$ is synthetically generated by (i) computing all displacement components at all nodes of the fine mesh for 100 time steps, (ii) interpolating them onto a coarser uniform mesh (i.e. cylindrical inclusion not included), consisting of $275,000$ tetrahedral elements ($50,000$ nodes), and (iii) polluting it with Gaussian noise with relative noise level $\delta = 0.01$. The total number of unknown moduli for this example is $550,000$. We note that reconstruction of moduli from interior data is the main focus of the active research area of biomechanical imaging (see e.g. [12] and references therein) .The value of ξ is selected according to δ via Morozov's discrepancy principle, i.e. choosing the largest value of ξ meeting the condition (wherein $(\boldsymbol{u}_k^{\star})_{0 \le k \le N}$ satisfy the optimality system (18)):

$$\sum_{k=1}^{N} \|\boldsymbol{u}_k^{\star} - \tilde{\boldsymbol{u}}_k\|^2_{L^2(\Omega_m)} \Big/ \sum_{k=1}^{N} \|\tilde{\boldsymbol{u}}_k\|^2_{L^2(\Omega_m)} \le \delta^2, \tag{21}$$

Figure 1 shows threshold plots of the reconstructed shear and bulk moduli, respectively, with the lower threshold set to 5 and 12, respectively, for plotting G and B. Either plot shows the inclusion location and geometry to be correctly identified (with some aberrations occurring near the top surface, which may result from measurements being less sensitive to moduli in those areas). Figure 2 shows cuts perpendicular to the x, y and z axes, going through the cube center.

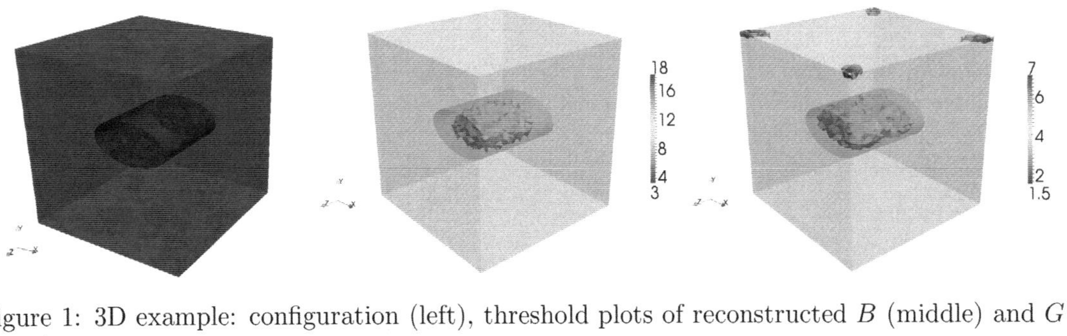

Figure 1: 3D example: configuration (left), threshold plots of reconstructed B (middle) and G (right)

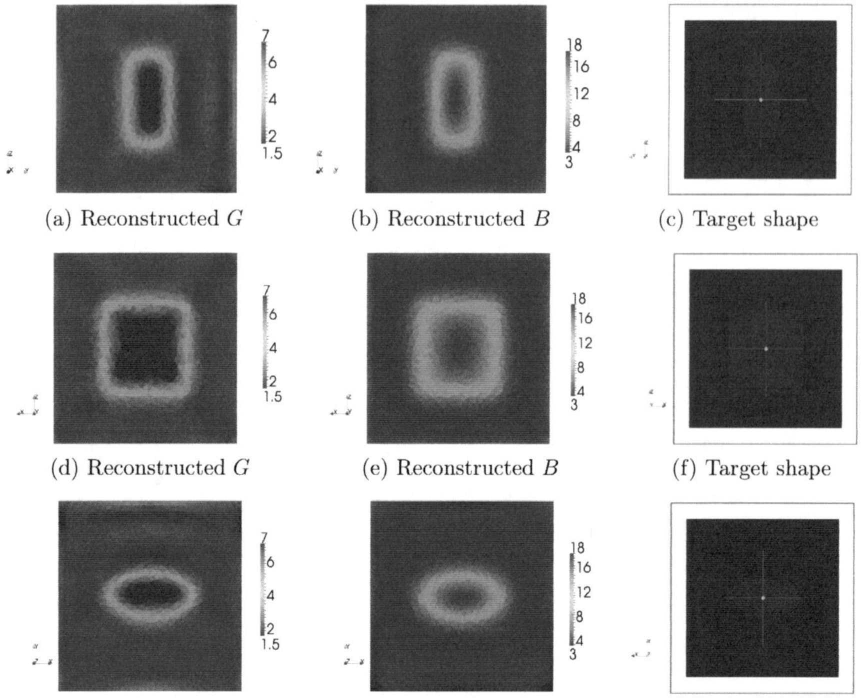

(a) Reconstructed G (b) Reconstructed B (c) Target shape

(d) Reconstructed G (e) Reconstructed B (f) Target shape

(g) Reconstructed G (h) Reconstructed B (i) Target shape

Figure 2: 3D example, cut views: x-plane (top row), y-plane (middle row), z-plane (bottom row)

References.

[1] Aguilo, M. A., Brigham, J. C., Aquino, W., Fatemi, M. An Inverse Problem Approach for Elasticity Imaging through Vibroacoustics. *IEEE Transactions on Medical Imaging*, pp. 1012–1021 (2010).

[2] Allix, O., Feissel, P., Nguyen, H. M. Identification strategy in the presence of corrupted measurements. *Eng. Comp.*, **22**:487–504 (2005).

[3] Banerjee, B., Walsh, T. F., Aquino, W., Bonnet, M. Large scale parameter estimation problems in frequency-domain elastodynamics using an error in constitutive equation functional. *Comp. Meth. Appl. Mech. Eng.*, **253**:60–72 (2013).

[4] Barthe, D., Deraemaeker, A., Ladevèze, P., Le Loch, S. Validation and updating of industrial models based on the constitutive relation error. *AIAA Journal*, **42**:1427–1434 (2004).

[5] Epanomeritakis, I., , Akcelik, V., Ghattas, O., Bielak, J. A Newton-CG method for large-scale three-dimensional elastic full-waveform seismic inversion. *Inverse Prob.*, **24**:034015 (2008).

[6] Feissel, P., Allix, O. Modified constitutive relation error identification strategy for transient dynamics with corrupted data: the elastic case. *Comp. Meth. Appl. Mech. Eng.*, **196**:1968–1983 (2006).

[7] Geymonat, G., Pagano, S. Identification of mechanical properties by displacement field measurement: a variational approach. *Meccanica*, **38**:535–545 (2003).

[8] Kohn, R.V., Lowe, B.D. A variational method for parameter identification. *Math. Mod. Num. Anal.*, **22**:293–315 (1988).

[9] Ladevèze, P., Leguillon, D. Error estimate procedure in the finite element method and applications. *SIAM J. Numer. Anal.*, **20**:485–509 (1983).

[10] Ladevèze, P., Nedjar, D., Reynier, M. Updating of finite element models using vibration tests. *AIAA Journal*, **32**:1485–1491 (1994).

[11] Métivier, L., Brossier, R., Virieux, J., Operto, S. Full waveform inversion and the truncated Newton method. *SIAM J. Sci. Comput.*, **35**:B401–B437 (2013).

[12] Sinkus, R., Tanter, M., Catheline, S., Lorenzen, J., Kuhl, C., Sondermann, E., Fink, M. Imaging anisotropic and viscous properties of breast tissue by magnetic resonance-elastography. *Magnetic Resonance in Medicine*, **53**:372–387 (2005).

[13] Walsh, T. F., Reese, G. M., Bhardwaj, M. K. Salinas: theory manual. Sandia report SAND2011-8272, http://dx.doi.org/10.2172/1031884 (2011).

Goal-oriented updating of mechanical models

L Chamoin[1], P Ladevèze[1] and J Waeytens[2]

[1]LMT-Cachan (ENS Cachan/CNRS), 61 Av. du Prés. Wilson, 94235 Cachan Cedex, France
[2]Paris-Est University, IFSTTAR, 14-20 Bd Newton, 77447 Marne-La-Vallée Cedex 2, France

E-mail: chamoin@lmt.ens-cachan.fr

Résumé. We introduce a goal-oriented procedure for the updating of mechanical models. It exploits as usual information coming from experimental data, but these are post-processed in a specific way in order to firstly update model parameters which are the most influent for the prediction of a given quantity of interest. The objective is thus to perform a partial model updating that enables to obtain an approximate value of the quantity of interest with sufficient accuracy and minimal model identification effort. The updating method uses the constitutive relation error framework, as well as duality and adjoint techniques, and defines dedicated cost functions. It leads to a convenient strategy that automatically selects the relevant parameter set to be updated. Performances of the approach are analyzed on examples involving linear elasticity and transient thermal models with possible noisy measurements.

1. Introduction

A major concern with mathematical models is their capability to represent a faithful abstraction of the real world. To address this issue and control the error between physical and mathematical models, model validation methods have been used for a long time [13]. In such methods, model parameters are identified or updated in order to minimize the discrepancy between numerical predictions and experimental measurements. The process leads to inverse problems which are usually ill-posed and require special care and regularization techniques in order to ensure solvability [3].
We focus here on the updating of Computational Mechanics models, in which a major component is the constitutive equation that describes the local behavior of the material [6]. We consider that only few localized measurements are available. Several procedures exist in this framework to identify parameters, both in a deterministic setting (minimization of cost functions associated with regularization techniques [3]) and in a stochastic setting (Bayesian inference [14]). The concept of Constitutive Relation Error (CRE) defines another model updating we use here. First introduced for dynamics models [10, 2, 7], this method was latter successfully used in many calibration applications with defects [5], uncertain measurements and behaviors [12], or corrupted measurements [1]. Recent applications of the method dealt with the updating of models used in bolted assemblies [8], or association with PGD reduced models to deal with real-time calibration of machining models [4]. The use of the CRE presents various advantages, such as the capacity to localize structural defects spatially, the robustness with respect to noisy measurements, and good convexity properties. In the CRE framework, reliable theoretical and experimental information (equilibrium, sensor position) are favored compared to other information (material behavior, sensor measures), and the hierarchical updating which is employed (only most erroneous zones

are corrected) directly provides for a regularization process.

In this work, we consider that the prediction target is only the value of a given output of the model (quantity of interest) that implicitly depends on model parameters and is relevant for design (local stress, maximal displacement,...). Therefore, if the quantity of interest is not very sensitive with respect to some parameters, there is probably no need to estimate these parameters with high accuracy. This would lead to a partial calibration of the model so as to ensure the quality of predicted quantities of interest with a minimal calibration effort. We thus define a goal-oriented version of updating methods performed using the CRE, focusing on the sensitivity of the considered quantity of interest with respect to parameters and measurements. We define new dedicated cost functions that lead to a convenient goal-oriented updating process, selecting automatically the relevant model parameters that need to be updated for the prediction of the quantity of interest. For the sake of simplicity, we present the new method and tools in the framework of linear elasticity models, whereas numerical experiments will also involve time-dependent problems.

2. Classical updating method based on the constitutive relation error

We consider a linear elastic body whose undeformed configuration occupies domain $\Omega \subset \mathbb{R}^d$, with boundary $\partial\Omega$. It is subjected to a given displacement field $\mathbf{u}_d$ on $\partial_1\Omega \subset \partial\Omega$, whereas a body force field $_\mathbf{d}$ and a traction force field $\mathbf{F}_d$ are imposed in Ω and on $\partial_2\Omega$, respectively, with $\partial_1\Omega \cup \partial_2\Omega = \partial\Omega$ and $\partial_1\Omega \cap \partial_2\Omega = \emptyset$. The corresponding model is governed by compatibility, equilibrium and constitutive equations which respectively read :

$$\varepsilon(\mathbf{u}) = \frac{1}{2}(\nabla\mathbf{u} + \nabla^T\mathbf{u}) \quad \text{in } \Omega \quad ; \quad \mathbf{u} = \mathbf{u}_d \quad \text{on } \partial_1\Omega \tag{1}$$

$$\mathbf{div}\,\sigma + _\mathbf{d} = \mathbf{0} \quad \text{in } \boldsymbol{\Omega} \quad ; \quad \sigma\mathbf{n} = \mathbf{F_d} \quad \text{on } \boldsymbol{\partial_2\Omega} \tag{2}$$

$$\sigma = \boldsymbol{K}\varepsilon(\mathbf{u}) \quad \text{in } \Omega \tag{3}$$

$\mathbf{u}$ is the displacement vector, ε the linearized strain tensor, σ the Cauchy stress tensor, and $\mathbf{n}$ the outward unit normal vector. $\boldsymbol{K}$ denotes the Hooke tensor, possibly heterogeneous, and we suppose it is described by a set $\mathbf{p}$ of parameters. We introduce classical sets $\mathcal{U} = \{\mathbf{v} \in [H^1(\Omega)]^d, \mathbf{v} \text{ verifies (1)}\}$ and $\mathcal{S} = \{\pi \in [L^2(\Omega)]^{d(d+1)/2}, \pi = \pi^T, \pi \text{ verifies (2)}\}$. Equations (1–3) constitute a direct problem that can be recast in the following weak form :

$$\text{Find } \mathbf{u}(\mathbf{p}) \in \mathcal{U} \text{ such that } \mathcal{A}(\mathbf{p},\mathbf{u},\mathbf{v}) = a(\mathbf{p},\mathbf{u},\mathbf{v}) - l(\mathbf{v}) = 0 \quad \forall \mathbf{v} \in \mathcal{U}_0 \tag{4}$$

where $\mathcal{U}_0$ denotes the vectorial space associated with $\mathcal{U}$.

The constitutive relation error (CRE) is a concept with strong mechanical content [11]. It defines an energy measure, denoted $\mathcal{E}$, of the distance between a given stress field π and another stress field obtained from a given displacement field $\mathbf{v}$ using (3) :

$$\mathcal{E}^2(\mathbf{p},\mathbf{v},\pi) = \frac{1}{2}\int_\Omega [\pi - \boldsymbol{K}(\mathbf{p})\varepsilon(\mathbf{v})] : \boldsymbol{K}^{-1}(\mathbf{p})\,[\pi - \boldsymbol{K}(\mathbf{p})\varepsilon(\mathbf{v})]\,\mathrm{d}\Omega \tag{5}$$

Its usefulness is explained by the fact that the solution $(\mathbf{u},\sigma)$ of the well-posed direct problem (1–3) is characterized by $(\mathbf{u},\sigma) = \text{argmin}_{(\mathbf{v},\pi)\in\mathcal{U}\times\mathcal{S}}\,\mathcal{E}(\mathbf{p},\mathbf{v},\pi)$. In recent model calibration formulations using the CRE, the cost (misfit) function $\mathcal{F}$ is constructed from $\mathcal{E}$; boundary conditions coming from experimental data are relaxed and imposed by penalization in a new definition of the constitutive relation error called *modified constitutive relation error* [10, 2, 7] and denoted $\mathcal{E}_m$:

$$\mathcal{E}_m^2(\mathbf{p},\mathbf{v},\pi) = \mathcal{E}^2(\mathbf{p},\mathbf{v},\pi) + \frac{1}{2}\frac{r}{1-r}||\mathbf{s}[\mathbf{v}] - \mathbf{s}_{obs}||^2_{L^2} \tag{6}$$

The two terms composing $\mathcal{E}_m$ can be respectively seen as measures of modeling error and measurement error. r is a scalar parameter that enables to modulate the influence of these terms. The philosophy of the modified constitutive relation error is thus to favor reliable theoretical and experimental information (equilibrium, sensor location) compared to other information (constitutive relation, sensor measurement). The inverse problem is defined as :

$$\mathbf{p}_{sol} = \underset{\mathbf{p}\in\mathcal{P}}{\operatorname{argmin}}\,\mathcal{F}(\mathbf{p}) \quad ; \quad \mathcal{F}(\mathbf{p}) = \min_{(\mathbf{v},\pi)\in\mathcal{U}\times\mathcal{S}} \mathcal{E}_m^2(\mathbf{p},\mathbf{v},\pi) \tag{7}$$

It leads to an iterative method, each iteration consisting of two minimizations steps :

- the first minimization step involved in (7), i.e. the computation of $\mathcal{F}(\mathbf{p})$ is called the *localization step*. This is a constrained minimization problem solved by means of a Lagrangian. Splitting the modeling error term of $\mathcal{F}(\mathbf{p})$ into contributions of each element p_i of $\mathbf{p}$ enables to localize the set of parameters that contribute the most to the error. Contributions of the measurement error enable to detect erroneous sensors;
- the second minimization step involved in (7), i.e. the computation of $\mathbf{p}_{sol}$ is called the *correction step*. In practice, $\mathcal{F}(\mathbf{p})$ is minimized with respect to parameters p_i chosen in the previous step. This is a nonlinear process that uses an optimization algorithm, such as the gradient method with optimal path.

The iterative process with localization and correction steps is in practice stopped when the cost function $\mathcal{F}(\mathbf{p})$ reaches a given tolerance value, without waiting for convergence. Furthermore, only parameters which are selected in the localization step are updated during the correction step (hierarchical updating). This naturally regularizes the inverse formulation.

3. Goal-oriented updating with the constitutive relation error

In the previous section, the model was updated globally with respect to experiments as a (modified) constitutive relation error, defined on the whole domain Ω and with the whole parameter set $\mathbf{p}$, was used as a cost function. Furthermore, parameters to be updated first were selected using contributions of this global error. Here, we aim at modifying the formulation to obtain an updating process which is dedicated to the prediction of given outputs of the model. We consider a quantity of interest Q which is supposed to be the goal of the computation. Performing an optimal calibration process with respect to Q means updating relevant parameters only. We introduce in the following cost functions which are dedicated to this purpose.

3.1. Goal-oriented cost function when the quantity of interest is not measured

We assume here that the spatial region in which Q is defined is not a measurement point. Keeping the philosophy and flexibility of the modified CRE, composed of modeling and measurements error terms, we introduce a new cost function, denoted $\mathcal{F}_Q(\mathbf{p})$, associated with the considered quantity of interest. It reads :

$$\mathcal{F}_Q(\mathbf{p}) = \min_{q\in\mathbb{R}}\left[\frac{1}{2}|q - Q_{mod}(\mathbf{p})|^2 + \frac{1}{2}\frac{r}{1-r}|q - Q_{obs}(\mathbf{p})|^2\right] \tag{8}$$

The modeling error term $\frac{1}{2}|q - Q_{mod}(\mathbf{p})|^2$ involves a value Q_{mod} of the quantity of interest defined from the model only, i.e. by means of state equation (4) :

$$Q_{mod}(\mathbf{p}) = Q(\mathbf{u}_1(\mathbf{p})) \quad ; \quad \mathcal{A}(\mathbf{p},\mathbf{u}_1,\mathbf{v}) = 0 \quad \forall\mathbf{v}\in\mathcal{U}_0 \tag{9}$$

The measurement error term $\frac{1}{2}|q - Q_{obs}(\mathbf{p})|^2$ involves a value Q_{obs} of the quantity of interest defined from an interpolation of measurements $\mathbf{s}_{obs}$, i.e. :

$$Q_{obs}(\mathbf{p}) = Q(\mathbf{u}_2(\mathbf{p})) \quad ; \quad (\mathbf{u}_2, \sigma_2) = \underset{(\mathbf{v},\pi)\in\mathcal{U}\times\mathcal{S}}{\text{argmin}} \; \mathcal{E}^2_m(\mathbf{p}, \mathbf{v}, \pi) \tag{10}$$

The definition of $\mathbf{u}_2$ thus involves $\mathbf{s}_{obs}$ and leads to a system $\overline{\mathbb{K}}(\mathbf{p})\mathbf{U}_2 = \overline{\mathbf{F}}$ after discretization. Therefore, the first step in the iterative updating process consists in computing $\mathcal{F}_Q(\mathbf{p})$. As it is a constrained minimization problem with fixed $\mathbf{p}$, we introduce the semi-discrete Lagrangian :

$$\mathcal{L}_Q(\mathbf{p}, q, \mathbf{u}_1, \mathbf{U}_2, \boldsymbol{\mu}, \boldsymbol{\Lambda}) = \frac{1}{2}|q - Q(\mathbf{u}_1)|^2 + \frac{1}{2}\frac{r}{1-r}|q - Q(\mathbf{U}_2)|^2 - \mathcal{A}(\mathbf{p}, \mathbf{u}_1, \boldsymbol{\mu}) - \boldsymbol{\Lambda}^T(\overline{\mathbb{K}}(\mathbf{p})\mathbf{U}_2 - \overline{\mathbf{F}}) \tag{11}$$

Finding the saddle-point of $\mathcal{L}_Q$ leads to verify, in particular, the following adjoint problems for all $(\delta\mathbf{u}_1, \delta\mathbf{U}_2) \in \mathcal{U}_0 \times \mathcal{U}_0^h$ and with $\beta = r[Q(\mathbf{u}_1) - Q(\mathbf{U}_2)]$:

$$\mathcal{A}(\mathbf{p}, \delta\mathbf{u}_1, \boldsymbol{\mu}) = \beta\, Q'(\mathbf{u}_1; \delta\mathbf{u}_1) \quad \forall \delta\mathbf{u}_1 \in \mathcal{U}_0 \quad ; \quad \boldsymbol{\Lambda}^T\overline{\mathbb{K}}(\mathbf{p})\delta\mathbf{U}_2 = -\beta\, Q'(\mathbf{U}_2; \delta\mathbf{U}_2) \quad \forall \delta\mathbf{U}_2 \in \mathcal{U}_0^h \tag{12}$$

In a second step, the gradient of $\mathcal{F}_Q^h(\mathbf{p})$ is computed using the adjoint state method and from fields $(q, \mathbf{u}_1, \mathbf{U}_2, \boldsymbol{\mu}, \boldsymbol{\Lambda})$ obtained in the previous step :

$$\mathrm{d}_\mathbf{p}\mathcal{F}_Q^h(\mathbf{p}; \delta\mathbf{p}) = \mathcal{L}'_{Q,\mathbf{p}}(\mathbf{p}, q, \mathbf{u}_1, \mathbf{U}_2, \boldsymbol{\mu}, \boldsymbol{\Lambda}; \delta\mathbf{p}) = -a'_\mathbf{p}(\mathbf{p}, \mathbf{u}_1, \boldsymbol{\mu}; \delta\mathbf{p}) - \boldsymbol{\Lambda}^T \mathrm{d}_\mathbf{p}\overline{\mathbb{K}}(\mathbf{p}; \delta_\mathbf{p})\mathbf{U}_2 \tag{13}$$

Elements of $\mathbf{p}$ associated with high components for the gradient of $\mathcal{F}_Q^h(\mathbf{p})$ are then selected to drive the nonlinear hierarchical minimization.

3.2. Goal-oriented cost function when the quantity of interest is measured

In this specific case, there is no interpolation of the data and therefore the regularization term involved in the previous cost function is missing. We propose a new cost function based on a local version of the modified CRE, i.e. a measure at point (or in the area) where the quantity of interest is defined. We denote by Q^* the dual quantity of Q (in the energy sense), and we write the local constitutive relation linking Q^* and Q under the form $Q^* = k(\mathbf{p})Q$. The new cost function thus reads :

$$\mathcal{F}_Q(\mathbf{p}) = \min_{(\mathbf{v},\pi)\, \in\, \mathcal{U}\times\mathcal{S}} \mathcal{E}^2_{m,loc}(\mathbf{p}, \mathbf{v}, \pi) \tag{14}$$

with

$$\mathcal{E}^2_{m,loc}(\mathbf{p}, \mathbf{v}, \pi) = \frac{1}{2}|Q^*(\pi) - k(\mathbf{p})Q(\mathbf{v})|^2 + \frac{1}{2}\frac{r}{1-r}|Q(\mathbf{v}) - Q_{obs}|^2 \tag{15}$$

The two terms in $\mathcal{E}^2_{m,loc}$ correspond to local modeling and measurement errors, respectively. Following a procedure similar to the one detailed in Section ??, we write the discrete version of (15) :

$$\mathcal{E}^{h2}_{m,loc}(\mathbf{p}, \mathbf{U}, \mathbf{V}) = \frac{1}{2}|Q^*(\mathbf{V}) - k(\mathbf{p})Q(\mathbf{U})|^2 + \frac{1}{2}\frac{r}{1-r}|Q(\mathbf{U}) - Q_{obs}|^2 \tag{16}$$

with $\mathbb{K}(\mathbf{p})\mathbf{V} = \mathbf{F}$ (equilibrium), and the computation of $\mathcal{F}_Q^h(\mathbf{p})$ is performed by searching the saddle-point of the following lagrangian $\mathcal{L}_Q(\mathbf{p}, \mathbf{U}, \mathbf{V}, \boldsymbol{\Lambda}) = \mathcal{E}^{h2}_{m,loc}(\mathbf{p}, \mathbf{U}, \mathbf{V}) - \boldsymbol{\Lambda}^T(\mathbb{K}(\mathbf{p})\mathbf{V} - \mathbf{F})$. This leads to solving, in particular, the following adjoint problem for all $\delta\mathbf{V} \in \mathcal{S}_0^h$:

$$Q^{*'}_\mathbf{u}(\mathbf{V}; \delta\mathbf{V})\,(Q^*(\mathbf{V}) - k(\mathbf{p})Q(\mathbf{U})) - \boldsymbol{\Lambda}^T\mathbb{K}(\mathbf{p})\delta\mathbf{V} = 0 \tag{17}$$

The evaluation of the gradient of $\mathcal{F}_Q^h(\mathbf{p})$ is then direct with the adjoint state method :

$$\mathrm{d}_\mathbf{p}\mathcal{F}^h(\mathbf{p}; \delta\mathbf{p}) = \mathcal{L}'_\mathbf{p}(\mathbf{p}, \mathbf{U}, \mathbf{V}, \boldsymbol{\Lambda}; \delta\mathbf{p}) \tag{18}$$

with $(\mathbf{U}, \mathbf{V}, \boldsymbol{\Lambda})$ solution of (17). Here again, elements of $\mathbf{p}$ associated with high components of the gradient are selected in order to minimize $\mathcal{F}_Q^h(\mathbf{p})$ with respect to these parameters only.

3.3. Remarks on the regularization process

In the goal-oriented updating method involving cost functions $\mathcal{F}_Q$ introduced in Sections 3.1 or 3.2, it is important to notice that an iterative two-step (localization, correction) strategy is conserved. However, the localization step is no more based on a splitting of the modeling error term and comparison between parameter contributions ($\mathcal{F}_Q$ is local in space) as used in Section 2, but on the selection of parameters that bring highest components to the gradient of $\mathcal{F}_Q(\mathbf{p})$. In practice, at each iteration of the method, the parameter (or the set of parameters) that brings the highest gradient component is corrected first. If this correction is not associated with a significant decrease of the cost function value, the parameter (or the set of parameters) that brings the second highest gradient component is corrected, and the process is stopped when a significant decrease of $\mathcal{F}_Q(\mathbf{p})$ is observed. This hierarchical strategy naturally regularizes the goal-oriented inverse formulation.

4. Numerical results

We consider a 2D concrete beam with a steel bar Γ (Figure 1). We divide the steel bar (resp. the lower concrete part Ω_c) into five subdomains Γ_j (resp. Ω_{c_j}), $j \in \{1, .., 5\}$ and consider piecewise constant steel bar cross-section and concrete Young modulus. We denote by S_{b_j} the steel bar cross-section in Γ_j, E_{cj} the concrete Young modulus in Ω_{c_j}, and k_1 (resp. k_2) the rigidity of support 1 (resp. support 2). We take $L = 30\ m$ and $H = 1\ m$. To update beam parameters

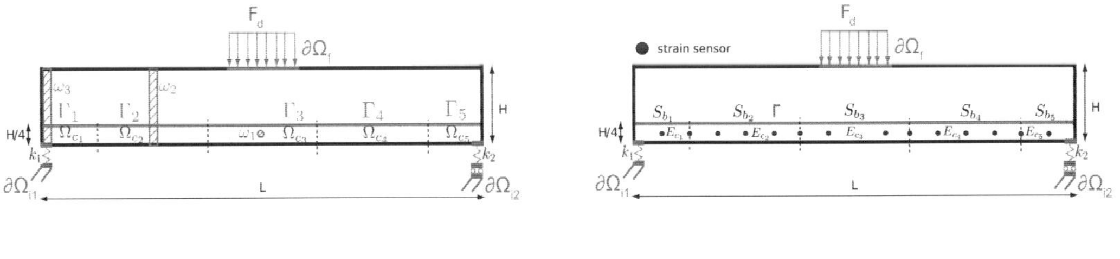

Figure 1. 2D concrete beam with a horizontal steel bar, with domains of interest ω_1, ω_2 and ω_3 (left), and instrumentation with 14 strain sensors (right).

$(\{S_{b_j}\}, \{E_{c_j}\}, k_1, k_2)$, a static loading $\mathbf{F}_d$ is applied and the longitudinal strain component ϵ_{xx} is measured at 14 points (see Figure 1). The sensor outputs are simulated numerically considering the direct model and the following reference beam parameters : $E_{c1}^{ref} = E_{c2}^{ref} = E_{c4}^{ref} = E_{c5}^{ref} = 40.10^9\ Pa$, $E_{c3}^{ref} = 30.10^9\ Pa$, $S_{b_1}^{ref} = S_{b_2}^{ref} = S_{b_3}^{ref} = S_{b_4}^{ref} = S_{b_5}^{ref} = 0.04\ m$, $k_1^{ref} = k_2^{ref} = 5.10^7\ N/m^3$. The only difference between reference and initial models is the value of the concrete Young modulus in subdomain Ω_{c3}.

The goal-oriented updating procedure is applied to three quantities of interest :

$$Q_1 = \int_{\omega_1} \epsilon_{xx} d\Omega \ ; \quad Q_2 = \int_{\omega_2} u_y d\Omega \ ; \quad Q_3 = \int_{\omega_3} u_y d\Omega \tag{19}$$

The associated zones of interest ω_1, ω_2 and ω_3 are represented in Figure 1.

Regarding the updating strategy, the highest component of the functional gradient is searched at each iteration and the associated parameter is updated. For each iteration of the goal-oriented technique, the updated parameter values and associated predicted values of the quantity of interest are represented in Figures 2 and 3. As regards quantities of interest Q_1 and Q_2, we observe that two iterations are sufficient to get an error less than 1% on the determination of the concrete Young modulus E_{c3} : we also emark that only parameter E_{c_3} has been updated using the goal-oriented approach. Quantity Q_3 being not sensitive to E_{c_3}, this parameter is not updated using the goal-oriented approach.

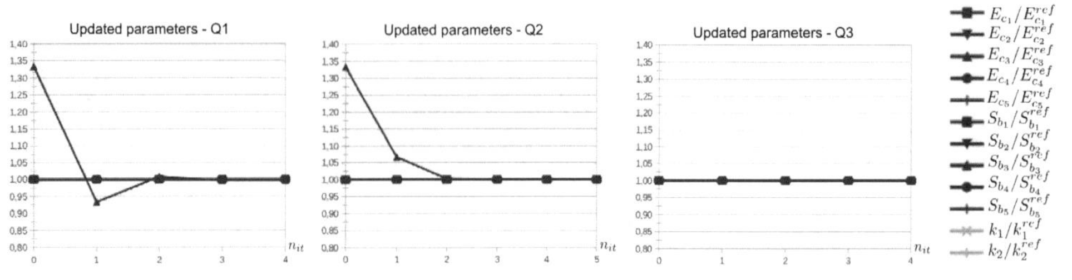

FIGURE 2. Updated beam parameters at each iteration of the updating goal-oriented approach for quantities of interest Q_1, Q_2, and Q_3.

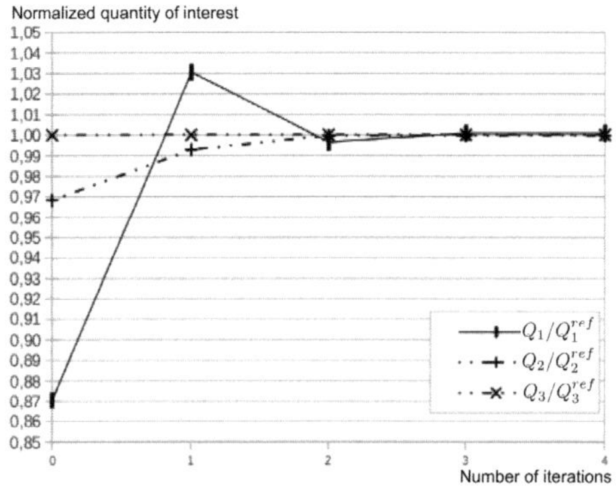

FIGURE 3. Ratio between the updated and the reference values of the quantity of interest at each iteration of the updating goal-oriented approach for Q_1, Q_2 and Q_3.

References

[1] Allix O, Feissel P and Nguyen H M 2005 Identification strategy in the presence of corrupted measurements *Eng. Comp.* **22** 487
[2] Bonnet M and Ben Abdallah J 1994 Structural parameter identification using nonlinear gaussian inversion *Inv. Prob. Eng. Mech.* 235
[3] Bonnet M and Constantinescu A 2005 Inverse problems in elasticity *Inv. Prob.* **21** R1
[4] Bouclier R, Louf F and Chamoin L 2013 Real-time validation of mechanical models coupling PGD and constitutive relation error *Comp. Mech* **52** 861
[5] Bui H D and Constantinescu A 2000 Spatial localization of the error of constitutive law for the identification of defects in elastic bodies. *Arch. Mech.* **52** 511
[6] Cailletaud G and Pilvin P 1994 Identification and inverse problems related to material behaviour *In H.D. Bui, M. Tanaka, and al., editors, Inv. Prob. Eng. Mech.* 79
[7] Chouaki A, Ladevèze P and Proslier L 1996 An updating of structural dynamic model with damping *Inv. Prob. Eng. : Theory and Practice* 335
[8] Gant F, Rouch P, Louf F and Champaney L 2011 Definition and updating of simplified models of joint stiffness *Int. J. Solids & Struct.* **48** 775
[9] Johansson H, Larsson F and Runesson K 2011 Application-specific error control for parameter identification problems *Int. J. Numer. Meth. Biomed. Eng.* **27** 608
[10] Ladevèze P, Nedjar D and Reynier M 1994 Updating of finite element models using vibration tests *AIAA Journal* **32** 1485
[11] Ladevèze P and Pelle J P 2004 *Mastering Calculations in Linear and Nonlinear Mechanics* Springer NY
[12] Ladevèze P, Puel G, Deraemaeker A and Romeuf T 2006 Validation of structural dynamics models containing uncertainties *Comp. Meth. Appl. Mech. Eng.* **195** 373
[13] Roache P J 1998 *Verification and Validation in Computational Science and Engineering* Hermosa publishers
[14] Tarantola A 2005 *Inverse problem theory and model parameter estimation* SIAM

Journal of Physics: Conference Series **542** (2014) 012005 doi:10.1088/1742-6596/542/1/012005

Physical Modeling of Microtubules Network

Pierre Allain, Charles Kervrann

Inria, Centre Rennes – Bretagne Atlantique, Campus Universitaire de Beaulieu, 35042 Rennes Cedex, France

E-mail: pierre.allain.t@gmail.com, charles.kervrann@inria.fr

Abstract. Microtubules (MT) are highly dynamic tubulin polymers that are involved in many cellular processes such as mitosis, intracellular cell organization and vesicular transport. Nevertheless, the modeling of cytoskeleton and MT dynamics based on physical properties is difficult to achieve. Using the Euler-Bernoulli beam theory, we propose to model the rigidity of microtubules on a physical basis using forces, mass and acceleration. In addition, we link microtubules growth and shrinkage to the presence of molecules (*e.g.* GTP-tubulin) in the cytosol. The overall model enables linking cytosol to microtubules dynamics in a constant state space thus allowing usage of data assimilation techniques.

1. Introduction

Microtubules (MTs) are essential for intracellular organization, cell division and cell morphogenesis. It is established that dynamic instability of MTs, with alternating phases of growth and shrinkage separated by catastrophe and rescue events, is regulated by microtubule-associated proteins (MAPs) [9, 4, 2, 8]. Nevertheless, understanding better the specific molecular interactions at microtubule growing ends recognized by MAPs and deciphering assembly mechanisms in the presence of guanosine triphosphate (GTP) is of major importance in cell biology and for cancer therapy.

Several simulation softwares to visualize the different models of microtubule dynamic instability have been already developed [3, 9], but essentially in 2D. In [5], the authors studied MT rigidity using microscopy imaging and physical laws which yielded mechanical properties measures. More recently, Portran et al. showed how MAP65, by modifying MT mechanical properties, regulate the formation of complex MT arrays [8]. In [6], the authors proposed a cytoskeleton simulation model based on derived kinematic laws of mechanics and suitable for fast simulation. However, most of the terms are somewhat empirically designed. Indeed, although this approach is effective for visual simulation, it does not take mass description into account and rather focus on cytoskeleton kinematics than its dynamics. More recently, computation tools that reproduce assembly of bundles made of MTs in the presence of MAP65 have been investigated to mimic in vivo MT bundles [9].

To get a better understanding of how geometrical constraints and interactions with the medium affect the MT organization and dynamics, we propose a new modeling framework which can be combined further with data assimilation techniques. Data assimilation techniques are essentially based on a controlled trade-off between observations of a phenomena and a model accounting for its likely dynamics. Such approach has been for example investigated in [1] for similar but simpler model.

In this paper, we focus on the modeling part of the general data assimilation framework. The model we propose links mechanically the cytoskeleton to the cytosol, and its resulting system allows investigation of an inverse problem. It is composed of two essential parts introduced in the next section: *i)* the Lagrangian dynamics of MTs (Section 2); *ii)* the Eulerian dynamics of the cytosol (Section 3).

2. Microtubule modeling

MTs are discretized by nodes connected by edges and are described either in 2D or 3D. The overall number of nodes $N = MP$ accounts for the size of the state space, while P accounts for the number of nodes for each microtubule, and M the number of microtubules. The set of nodes carries different properties such as positions $\mathbf{y} \in \mathbb{R}^{Nd}$, $d \in \{2, 3\}$, and velocities $\mathbf{u} \in \mathbb{R}^{Nd}$ that can be read for node $\mathbf{p}_i$ respectively as $y_i \in \mathbb{R}^d$ and $u_i \in \mathbb{R}^d$. These properties allows building the following Newtonian equations:

$$\frac{\partial y_i}{\partial t} = u_i \text{ and } \frac{\partial u_i}{\partial t} = \frac{1}{\mathrm{m}} \sum_f \mathbb{F}_{f,i} \tag{1}$$

where m accounts for the homogeneous mass of nodes, and $\mathbb{F}_f$ is a force applied to the nodes. In what follows, we will detail the forces of the model and needed expansions of the system to manage specific behaviors such as rigidity.

2.1. Microtubule rigidity

Using the Euler-Bernoulli beam theory, we model the rigidity of MTs on a physical basis. This mechanism relies on a difference between the shape of a given material and its neutral state. In this purpose, we introduce the neutral position state $\mathbf{z} \in \mathbb{R}^{Nd}$ accounting for the position of nodes not submitted to any forces.

2.1.1. Elasticity MTs are very rigid material. In [6], MTs are considered as non-elastic by using reshaping algorithm. Although this technique can be justified in practice, it involves an implicit integration potentially expensive from a calculation point of view regarding the number of nodes used for discretization. In order to keep a simple explicit integration, we prefer to consider the MTs as elastic with a high valued spring coefficient. As a counterpart, integration time-step has to be small. By defining $c^{\mathbf{y}}_{ij} = y_j - y_i$, $d^{\mathbf{y}}_{ij} = \|y_j - y_i\|$ and $e^{\mathbf{y}}_{ij} = c^{\mathbf{y}}_{ij}/d^{\mathbf{y}}_{ij}$ the elasticity force reads for each node $\mathbf{p}_i$:

$$\mathbb{F}_{\text{elasticity},i} = sE \sum_{j \leftrightarrow i}^{N} (d^{\mathbf{y}}_{ij} - d^{\mathbf{z}}_{ij}) e^{\mathbf{y}}_{ij} \tag{2}$$

where s denotes the section of MTs, E is Young modulus and $j \leftrightarrow i$ means particle $\mathbf{p}_j$ is connected to $\mathbf{p}_i$. We suppose MTs to be hollow, therefore we can write: $s = \pi(r^2 - r_0^2)$, with r and r_0 being respectively their outer and inner radius.

2.1.2. Bending Flexibility of MTs is the more tangible mechanism due to rigidity. Before expressing the forces applied on nodes, we detail how angular variation along MTs yields torque. For any node $\mathbf{p}_i$ connected to nodes $\mathbf{p}_j$ and $\mathbf{p}_k$, the torque is oriented in the direction:

$$\omega^{\mathbf{y}}_{jik} = \frac{c^{\mathbf{y}}_{ij} \times c^{\mathbf{y}}_{ik}}{\left\| c^{\mathbf{y}}_{ij} \times c^{\mathbf{y}}_{ik} \right\|}, \tag{3}$$

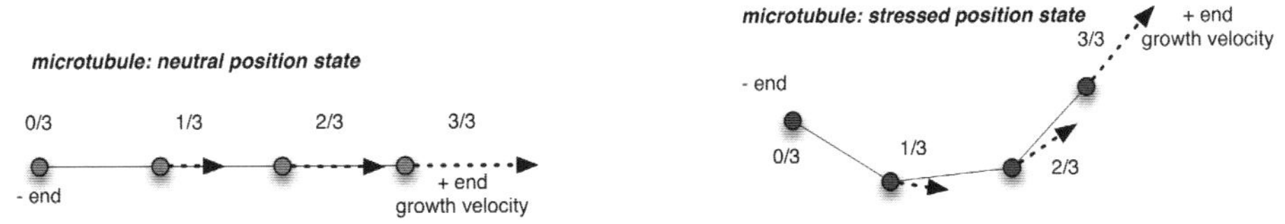

Figure 1. Growth velocity of microtubule: left: neutral state; right: position state.

except in the case y_i, y_j and y_k are aligned. We introduce $\mathbb{T}_{\text{bending}} \in \mathbb{R}^{Nd}$, the torque produced by angular variations along MTs, reading:

$$\mathbb{T}_{\text{bending},i} = IE \sum_{j \leftrightarrow i}^{N} \sum_{\substack{k \leftrightarrow i \\ k \neq j}}^{N} \left(\gamma_{jik}^{\mathbf{y}} - \gamma_{jik}^{\mathbf{z}} \right) \omega_{jik}^{\mathbf{y}} \quad \text{with} \quad \gamma_{jik}^{\mathbf{y}} = \frac{\|-y_j + 2y_i - y_k\|}{\frac{1}{4}\|y_j - y_k\|^2} \tag{4}$$

where $I = \frac{\pi}{4}(r^4 - r_0^4)$ is the second moment of area of MTs. Forces produced by torque reads:

$$\mathbb{F}_{\text{bending},i} = -\sum_{j \leftrightarrow i}^{N} \frac{c_{ij}^{\mathbf{y}}}{d_{ij}^{\mathbf{y}}} \times \mathbb{T}_{\text{bending},j}. \tag{5}$$

For $d = 3$, a torsion component exists, but we ignore it because of the edge modeling of MTs.

2.2. Collision

The scale difference between microtubule radius and cytosol domain size ($\simeq 10^3$ factor) yields numerical difficulties for collision management. A small time step is needed to accurately check whether two microtubules are colliding or not. For simulation ease, the collision is rather modeled as closeness preventing. This kind of modeling yields smooth collision management and easy to tune parameters. The resulting repulsion force is described as follows:

$$\mathbb{F}_{\text{repulsion},i} = \sum_{j, MT(\mathbf{p}_i) \neq MT(\mathbf{p}_j)}^{N} f_{\text{rep}}\, e^{-\frac{d_{ij}^{\mathbf{y}} - 2r}{d_{\text{rep}}}} e_{ij}^{\mathbf{y}}, \tag{6}$$

where f_{rep} is a reference force and d_{rep} is a reference distance of repulsion. Ideally, for perfect collision management with supposedly hard material, d_{rep} should tend to 0. Instead, we parameter this reference distance accordingly to MT thickness by $d_{\text{rep}} = \alpha r$.

2.3. Microtubule growth and shrinkage

In our approach, the number of nodes modeling MTs is constant. The main advantage is that the state space of the model keeps constant too. But as counterpart, managing the extension of curvilinear forms without modifying their shape and with the same discretization can be tedious. We suppose growth/shrinkage velocity to be directed same as microtubule plus-end edge. Considering the neutral position state $\mathbf{z}$ of microtubule, as in Fig. 1 (left), intensity of growth velocity is simply projected on nodes with a decreasing ratio such as:

$$v_{\text{growth},m,j} = \frac{j-1}{P-1} v_{\text{growth},m,P}, \tag{7}$$

where j is the node index inside microtubule MT_m, and $v_{\text{growth},m,P}$ is the growth velocity at its plus-end. Such ratio makes the assumption that distance between attached nodes is constant, which is reasonable due to the strong spring coefficient used in the experiments. As a consequence of edge-directed growing velocity, if MTs neutral positions are aligned at t_0, they will stay so. In this case, the component $\gamma_{jik}^{\mathbf{z}}$ in (4) is null.

Regarding the real microtubule position state $\mathbf{y}$, growth velocity projection applies to u_{growth} almost the same way. But as a difference, the direction of inner nodes velocity varies along microtubule length. For each node, the direction is set as the tangent of the two connected edges, while the same ratio as the neutral state is applied for intensity (see Fig. 1 (right)).

Table 1. Model parameters for MT dynamic simulation

r	r_0	E	f_{rep}	α	η	β	λ	ω_{growth}	ω_{shrink}	T_{ref}	T_{min}	T_{max}
15	12	1	10^{-8}	4	10^{-3}	$2.5\ 10^{-11}$	10^{-11}	10^{-16}	$2\ 10^{-16}$	1	0.8	1.2
nm	nm	GPa	N		kg.(m.s)$^{-1}$	m^2.s^{-1}	t.m^{-1}	t.s^{-1}	t.s^{-1}	t.m^{-d}	t.m^{-d}	t.m^{-d}

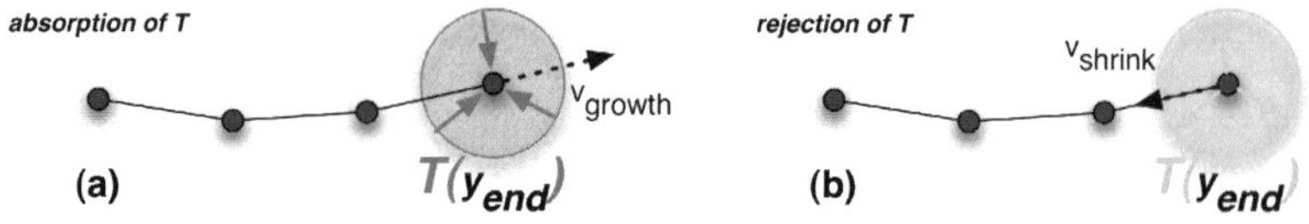

Figure 2. Growth interaction between MTs and cytosol. *a)* Absorption of "pseudo-tubulin" in cytosol at growing microtubule plus-end, *b)* rejection of "pseudo-tubulin" for shrinking microtubule.

Microtubules index space $\{j, m\}$ is linked to global index space $\{i\}$ thanks to the relation: $i = h(m, j) = Pm + j$. That allows us to express growth velocity in the global nodes space such as: $v_{\text{growth},i} = v_{\text{growth},h(m,j)}$ and $u_{\text{growth},i} = u_{\text{growth},h(m,j)}$. We shall see later how the intensity of growth velocity at plus-end node is determined.

2.4. Fluid dragging

MT movements also depend on cytosol dynamics. Various mechanical properties such as pressure or velocity can produce different effects on MT spatial evolution. We only use inner cytosol velocity $\mathbf{u}(\mathbf{x}, t)$ for viscous dragging of MTs, with $\mathbf{x}$ being the position in the cytosol space. Because of fluid-structure interactions, determining accurate drag forces is a tedious task calling for advanced computing fluid dynamics techniques. Fineness of MTs compared to cytosol spatial dimensions and slow velocities yielding a low Reynolds number allows one to make assumption the drag force is simply proportional to fluid velocity. We suppose the force proportional to the dynamic viscosity of cytosol and to the diameter of MTs. The viscous drag force is therefore defined as: $\mathbb{F}_{\text{drag},i} = 2r\eta\mathbf{u}(y_i)$.

3. Cytosol modeling

We propose to link MT growth and shrinkage to the presence of molecules in the cytosol. The growth velocity of MTs depends upon soluble tubulin concentration available for polymerization and GTP-tubulin association and dissociation rates [9, 4, 2]. The shrinkage velocity is known to be independent of tubulin concentration and is characterized only by the dissociation rate of guanosine diphosphate (GDP) tubulin from the depolymerizing end. In our modeling approach, we do not refer MT growth explicitly to the presence of GTP-tubulin since MT dynamics depend also on numerous properties of the cytosol. In what follows, we only refer to the concentration of a meta-quantity responsible for MT growth and shrinkage. This concentration can be also viewed as a *growth potential* whose distribution indicates locations of available growth energy. Due to the important role of GTP-tubulin molecules in microtubule polymerization, we call it "pseudo-tubulin" concentration written $T(\mathbf{x}, t)$ with the attached unit $[T] = t.m^{-3}$.

3.1. "Pseudo-tubulin" modeling

We propose to model evolution of "pseudo-tubulin" using a classical advection-diffusion model with an additional term A accounting for microtubules absorption and release:

$$\frac{\partial T}{\partial t} + \mathbf{u}\nabla T = \beta \Delta T + A(\mathbf{x}) \quad \text{and} \quad T(\mathbf{x}, t_0) = T_{\text{ref}} \tag{8}$$

with β being the diffusion coefficient of "pseudo-tubulin" and T_{ref} the reference value of T set to $1\,t.m^{-3}$. We suppose the absorption term to depend on MT growth velocity at their extremity, as illustrated in Fig. 2.

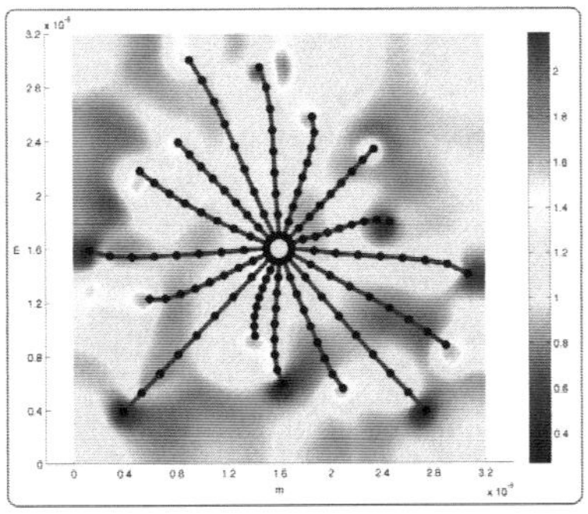

Figure 3. Simulation of a 2D radial microtubule network. The results show growing and shrinking phases yielding inhomogeneous "pseudo-tubulin" concentration in the cytosol. MTs are bended according to fluid forces.

Theoretically the absorption is only localized on this specific position, but due to spatial discretization of the cytosol on a grid, it is necessary to use a kernel convolution, and finally the term reads:

$$A(\mathbf{x}) = \sum_{m}^{M} w_m \boldsymbol{G}_\sigma(\mathbf{x}) * \lambda \left\| v_{\text{growth},m} \right\| \delta_{y_{m,P}}, \tag{9}$$

where w_m values -1 if microtubule m is growing, 1 if shrinking. Also, $\boldsymbol{G}_\sigma$ is the gaussian kernel of standard deviation σ, $*$ is the convolution operator, δ is the Dirac function, and λ stands for the microtubule linear density of "pseudo-tubulin" such as $[\lambda] = t.m^{-1}$. In case "pseudo-tubulin" is considered as α-tubulin and β-tubulin, it is possible to parameter this coefficient with a known value according to their spatial configuration shaping microtubule structure. Otherwise, if T is considered as growing energy, λ can be related to linear energy capacity. Since $\boldsymbol{G}_\sigma$ has to be as close as possible to de Dirac function, σ is set equal to the lowest significant value, being size of grid meshes.

3.2. Growth functional

We propose to link microtubules both growth and shrinking occurrence and intensity to "pseudo-tubulin" concentration T. The intensity of the growth velocity at microtubule plus-end is for instance set as:

$$\left\| v_{\text{growth},m,P} \right\| = \frac{\omega}{\lambda}\, g(T(y_{m,P})), \tag{10}$$

with ω standing for the microtubule "pseudo-tubulin" frequency and $g(T)$ is the intensity function. Their expression changes wether microtubule is growing or shrinking and reads as: $\omega = \omega_{\text{growth}}$ and $g(T) = \exp((T - T_{\text{ref}})/\Delta T)$ if growing and $\omega = \omega_{\text{shrink}}$ and $g(T) = \exp((T_{\text{ref}} - T)/\Delta T)$ if shrinking. Function $g(T)$ uses a "pseudo-tubulin" amplitude ΔT as parameter set as following: $\Delta T = T_{\max} - T_{\min}$ and $2\,T_{\text{ref}} = T_{\max} + T_{\min}$.

We make the assumption that for any growing microtubule m, whenever $T(y_{m,P})$ goes below $T_{\min}$, the available growing energy is not sufficient anymore to sustain the growing process. Therefore, the microtubule switch to a catastrophe phase. Oppositely, when microtubule is shrinking, whenever $T(y_{m,P})$ goes above $T_{\max}$, the available growing energy is sufficient to start again the growing process.

4. Final system and experiments

In the end, we obtain a system modeling MT rigidity, growth and interaction with cytosol. The space state describing this system is constant, which is a great advantage for analysis techniques. Finally, the overall system reads:

$$\mathbf{X} = \begin{bmatrix} \mathbf{z} \\ \mathbf{y} \\ \mathbf{u} \\ T \end{bmatrix}, \ \mathbb{M}(\mathbf{X}) = \begin{bmatrix} -\mathbf{v}_{\text{growth}} \\ -(\mathbf{u} + \mathbf{u}_{\text{growth}}) \\ -\frac{1}{\text{m}}\left(\mathbb{F}_{\text{elasticity}} + \mathbb{F}_{\text{bending}} + \mathbb{F}_{\text{repulsion}} + \mathbb{F}_{\text{drag}}\right) \\ \mathbf{u}\nabla T - \beta\Delta T - A(\mathbf{x}) \end{bmatrix}.$$

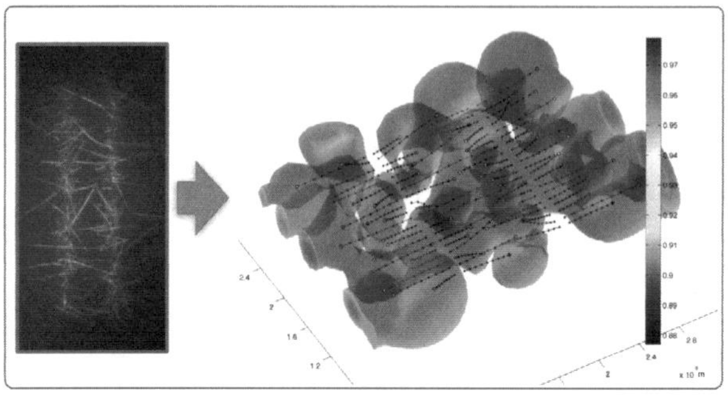

Figure 4. 3D simulation of MT nucleation and growth that mimics MT dynamics seeded onto a two vertical bar-shaped fibronectin pattern and observed in TIRFM (courtesy of iRTSV/LPCV/PCM CEA-Grenoble) [7].

In our experiments, velocity fields $\mathbf{u}(\mathbf{x}, t)$ are obtained by simulation of the incompressible Navier-Stokes equations, initialized by null divergence and random curl field. At initialization, "pseudo-tubulin" is randomly distributed around T_{ref}. Hence simulation is able to provide the expected variability. Simulation grid meshes size is set to 1 μm, while integration is performed using fourth order Runge-Kutta scheme with a 2 ms time-step. Results are presented in Figs. 3-4. One can observe our model yields appealing information on cytosol properties. In Fig. 4), simulations have been performed to mimic MT nucleation using MT microseeds adsorbed on a micropatterned glass substrate and observed in TIRF microscopy [7]. This approach is actually used to investigate MT functions using geometrically controlled MT networks in vitro in cell-free assay. Simulation parameters in Table 1 are chosen in agreement with published data.

5. Conclusion and perspectives

In vivo and in vitro, the MTs form networks resulting from their interplay with other protein partners [4]. The role of mechanical and/or geometrical constraints in the network organization is also important in that concern. Nevertheless, it is difficult to answer simple questions about the mechanics at the level of the whole network. In this paper, we have described a flexible mathematical model for representing MT network dynamics and molecular interactions. This model is able to manage microtubule rigidity, growth and interactions with cytosol. It addresses questions about the mechanisms and the forces developed at the level of MT networks and it has been designed so as to be compliant to analysis techniques based on data assimilation. To go further, we will derive this model in order to apply variational assimilation principles inspired of geophysical analysis. We plan also to combine simulation/data assimilation methods, in vitro experiments to constraint MT dynamics on patterns with specific geometry.

6. References

[1] P. Allain, N. Courty, and T. Corpetti. Particle swarm control. *Optimal Control Applications and Methods*, pages – to appear, 2013.

[2] F. Coquelle, S. Blestel, C. Heichette, I. Arnal, C. Kervrann, and D. Chrétien. Cryo-electron tomography of MTs assembled in vitro from purified components. *Meth. Mol. Biol.*, 777:193–208, 2011.

[3] A. Dimitrov, M. Quesnoit, S. Moutel, I. Cantaloube, C. Pous, and F. Perez. Detection of GTP-Tubulin conformation in vivo reveals a role for GTP remnants in MT rescues. *Science*, 322:1353–1356, 2008.

[4] M.K. Gardner, B.D. Charlebois, I.M. Janosi, J. Howard, A.J. Hunt, and D.J. Odde. Rapid MT self-assembly kinetics. *Cell*, 146(4):582–592, 2011.

[5] F. Gittes, B. Mickey, J. Nettleton, and J. Howard. Flexural rigidity of microtubules and actin filaments measured from thermal fluctuations in shape. *The Journal Cell Biol.*, 120(4):923–934, 02 1993.

[6] F. Nedelec and D. Foethke. Collective langevin dynamics of flexible cytoskeletal fibers. *New Journal of Physics*, 9(11):427, 2007.

[7] D. Portran, J. Gaillard, M. Vantard, and M. Thery. Quantification of map activity on geometrically controlled microtubule networks. *Cytoskeleton*, 70(1):12–23, 2013.

[8] D. Portran, M. Zoccoler, J. Gaillard, V. Stoppin-Mellet, E. Neumann, I. Arnal, J.L. Martiel, and M. Vantard. Map65/ase1 increase microtubule flexibility. *Mol. Biol. Cell*, 24(12):1964–1973, 2013.

[9] V. Stoppin-Mellet, V. Fache, D. Portran, J.-L. Martiel, and M. Vantard. MAP65 coordinate MT growth during bundle formation. *PLoS One*, 8(2):e56808, 2013.

An alternating minimization method for blind deconvolution from Poisson data

Marco Prato[1], Andrea La Camera[2] and Silvia Bonettini[3]

[1] Dipartimento di Scienze Fisiche, Informatiche e Matematiche, Università di Modena e Reggio Emilia, Via Campi 213/b, 41125 Modena, Italy

[2] Dipartimento di Informatica, Bioingegneria, Robotica e Ingegneria dei Sistemi, Università di Genova, Via Dodecaneso 35, 16145 Genova, Italy

[3] Dipartimento di Matematica e Informatica, Università di Ferrara, Via Saragat 1, 44122 Ferrara, Italy

E-mail: marco.prato@unimore.it, andrea.lacamera@unige.it, silvia.bonettini@unife.it

Abstract. Blind deconvolution is a particularly challenging inverse problem since information on both the desired target and the acquisition system have to be inferred from the measured data. When the collected data are affected by Poisson noise, this problem is typically addressed by the minimization of the Kullback-Leibler divergence, in which the unknowns are sought in particular feasible sets depending on the a priori information provided by the specific application. If these sets are separated, then the resulting constrained minimization problem can be addressed with an inexact alternating strategy. In this paper we apply this optimization tool to the problem of reconstructing astronomical images from adaptive optics systems, and we show that the proposed approach succeeds in providing very good results in the blind deconvolution of non-dense stellar clusters.

1. Blind deconvolution from Poisson data

Several acquisition systems provide the information as the convolution of an unknown distribution with a hardware-dependent kernel which transforms the target input in the measured output. A common instance of deconvolution problem is the image deblurring, in which the target image is distorted by a point spread function (PSF) whose features depend on the acquisition system and the environmental occurrences [1, 2]. If the PSF is not completely known, it has to be estimated within the deconvolution process together with the unknown distribution and the reconstruction problem increases its ill-posedness due to the large number of possible solutions, most of which have no physical meaning. This problem is known in the literature as blind deconvolution, and a key requirement for its solution is the availability of suitable a priori information. In this paper we address the blind deconvolution problem of reconstructing a blurred astronomical image acquired by a ground-based telescope when the PSF is unknown. Due to the Poisson nature of the noise introduced in the recording process of the image, a maximum likelihood approach leads to the reformulation of the imaging problem as the constrained minimization of the Kullback-Leilber (or Csiszár I- [3]) divergence

$$KL(\boldsymbol{g}, \boldsymbol{h} * \boldsymbol{f} + \boldsymbol{b}) = \sum_{i \in S} \left\{ g_i \log \frac{g_i}{(\boldsymbol{h} * \boldsymbol{f})_i + b_i} + (\boldsymbol{h} * \boldsymbol{f})_i + b_i - g_i \right\}, \tag{1}$$

where $\boldsymbol{g}$ is the measured image, $\boldsymbol{h}$ the PSF, $\boldsymbol{f}$ the target image, $\boldsymbol{b}$ a known background radiation, S the set of multi-indices $\boldsymbol{i} = (i_1, i_2)$ and $\boldsymbol{h} * \boldsymbol{f}$ denotes the convolution product. As concerns the feasible sets for $\boldsymbol{f}$ and $\boldsymbol{h}$, we consider non-negativity and flux conservation for the target image, while for the PSF we impose non-negativity, normalization to 1 and an upper bound deduced by the so-called Strehl ratio (SR), which is the ratio of peak diffraction intensity of an aberrated versus perfect waveform and can be estimated from the telescope and image features (see e.g. [4]). As a result of these choices, the mathematical formulation of the minimization problem becomes

$$\begin{aligned} &\min \;\; KL(\boldsymbol{g}, \boldsymbol{h} * \boldsymbol{f} + \boldsymbol{b}), \\ &\text{s.t.} \;\; \boldsymbol{f} \geq 0 \,, \sum_{\boldsymbol{i} \in S} \boldsymbol{f}_{\boldsymbol{i}} = c \;;\; 0 \leq \boldsymbol{h} \leq s \,, \sum_{\boldsymbol{i} \in S} \boldsymbol{h}_{\boldsymbol{i}} = 1 \end{aligned} \tag{2}$$

where c is the background-subtracted flux of the detected image and s is the upper bound derived from the SR.

2. Alternating minimization

Problem (2) is a hard task for an optimization method since the objective function is convex if restricted to $\boldsymbol{f}$ or $\boldsymbol{h}$, but is nonconvex with respect to $(\boldsymbol{f}, \boldsymbol{h})$ [5], thus admitting multiple stationary points. Due to the separability and the simplicity of the constraints, here we adopt an inexact alternating minimization (IAM) strategy [6, 7, 8], in which at each (outer) iteration $k \geq 1$ the two subproblems

$$\begin{aligned} &\operatorname{argmin} \;\; KL(\boldsymbol{g}, \boldsymbol{h}^{(k-1)} * \boldsymbol{f} + \boldsymbol{b}) \\ &\text{s.t.} \;\; \boldsymbol{f} \geq 0 \,, \sum_{\boldsymbol{i} \in S} \boldsymbol{f}_{\boldsymbol{i}} = c \end{aligned} \tag{3}$$

$$\begin{aligned} &\operatorname{argmin} \;\; KL(\boldsymbol{g}, \boldsymbol{h} * \boldsymbol{f}^{(k)} + \boldsymbol{b}) \\ &\text{s.t.} \;\; 0 \leq \boldsymbol{h} \leq s \,, \sum_{\boldsymbol{i} \in S} \boldsymbol{h}_{\boldsymbol{i}} = 1 \end{aligned} \tag{4}$$

are sequentially solved inexactly by means of a finite numbers $N_f^{(k)} \leq N_f$ and $N_h^{(k)} \leq N_h$ (with N_f and N_h arbitrarily fixed) of (inner) iterations of the scaled gradient projection (SGP) method [9]. The convergence of the resulting scheme has been proved in Theorem 4.2 of [6]:

Theorem 2.1 *For any choice of* (N_f, N_h)*, every limit point of the sequence* $(\boldsymbol{f}^{(k)}, \boldsymbol{h}^{(k)})$ *generated by applying* $N_f^{(k)} \leq N_f$ *and* $N_h^{(k)} \leq N_h$ *SGP iterations to* (3) *and* (4)*, respectively, is a stationary point for problem* (2).

Since SGP has been described in several papers concerning deblurring and denoising problems [9, 10, 11, 12, 13, 14], here we do not address the steps of the algorithm and its details. A complete and recent description can be found in [15], while the routines in MATLAB and Interactive Data Language (IDL) implementing SGP can be freely downloaded by the website http://www.unife.it/prisma/software. The result in Theorem 2.1 is the usual convergence property proved for a general first-order method in absence of further convexity hypotheses of the objective function. We remark that, thanks to the constraints we adopted, the sequence $(\boldsymbol{f}^{(k)}, \boldsymbol{h}^{(k)})$ generated by our method is bounded and therefore the existence of limit points is assured, even if there is no practical way to demonstrate if these points are minimizers.

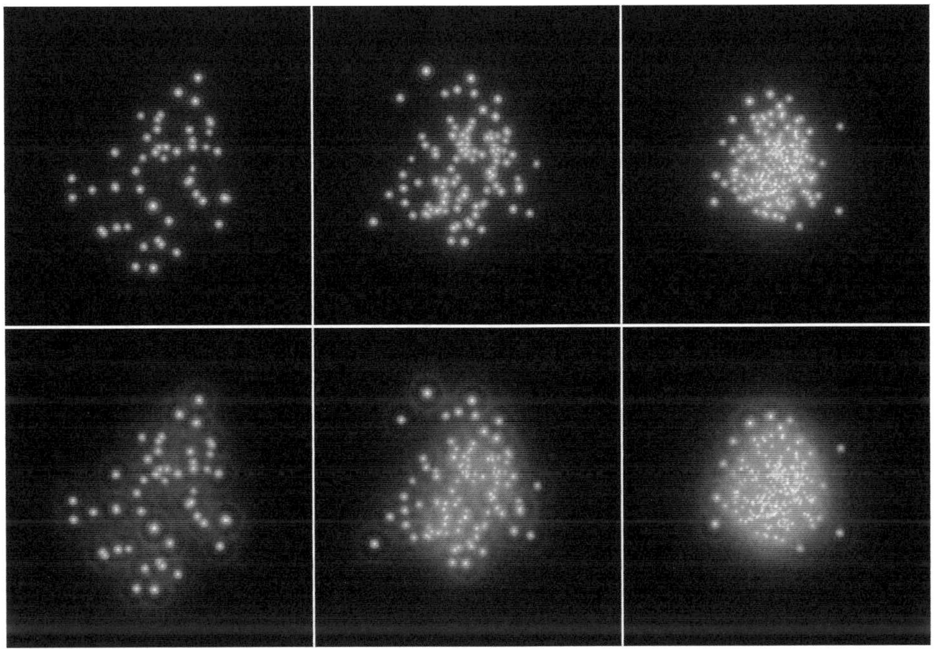

Figure 1. Blurred and noisy images of the star clusters SC50 (left panels), SC100 (central panels) and SC200 (right panels) convolved with the PSFs of LBT with SR = 0.81 (top panels) and 0.60 (bottom panels). All images are displayed in log scale.

3. Numerical experiments

In this section we test the effectiveness of the IAM procedure in an astronomical blind deconvolution problem. In particular, we used simulations of a single mirror of the Large Binocular Telescope (LBT) and its First Light Adaptive Optics (FLAO) system [16], which provides SR up to 0.9 in K-band.

3.1. Creating the simulations

We chose two PSFs with SR = 0.81 and 0.60, respectively, while for the astronomical objects we restricted our analysis to the 512×512 images of three star clusters, in which the positions of the stars have been computed following a Gaussian distribution around the center of the image and their magnitudes are randomly distributed around $m = 16$ with a standard deviation of about 0.4. The number of stars in the two images has been set equal to 50, 100 and 200, respectively. The three targets will be denoted in the following with SC50, SC100 and SC200. We convolved the three objects with the two PSFs, thus obtaining six noise-free images, to which we added a background in K-band, corresponding to about 13.5 mag arcsec^{-2}.
The resulting images have been corrupted with Poisson and additive Gaussian noise (read out noise - RON). In order to avoid saturation of the detector, the images have been obtained by co-adding 30 frames. It follows that the variance of the RON is $30\sigma^2$, thus corresponding to the RON of 30 frames; we set $\sigma = 10\ e^-/px$. According to the approach proposed in [17], RON compensation is obtained in the deconvolution algorithms by adding the constant $30\sigma^2 = 3000$ to the image and the background. The final images are shown in figure 1.

Table 1. Reconstruction errors for the SC50, SC100 and SC200 star cluster models and the two SRs. In the third column the initialization $\boldsymbol{h}^{(0)}$ is given (A is the autocorrelation of the ideal PSF while C is the ideal PSF plus a constant selected for satisfying the SR constraint - see the text). In the subsequent columns, we give the value of the MARE defined in Eq. (5) and the RMSE for the reconstructed PSF. The values of the objective function FOBJ are also reported.

Star Cluster	SR	Init.	MARE	RMSE	FOBJ
SC50	0.81	A	0.08%	1.05%	9.99e+004
		C	0.07%	2.61%	1.02e+005
	0.60	A	0.03%	2.90%	1.05e+005
		C	0.04%	3.03%	1.00e+005
SC100	0.81	A	0.06%	0.86%	9.64e+004
		C	0.05%	3.25%	9.93e+004
	0.60	A	0.02%	4.07%	1.14e+005
		C	0.03%	3.75%	1.00e+005
SC200	0.81	A	0.06%	0.76%	9.30e+004
		C	0.03%	3.62%	9.67e+004
	0.60	A	0.02%	4.63%	1.31e+005
		C	0.02%	4.32%	1.02e+005

3.2. Initialization and iteration numbers

A key point of the procedure is a good choice of the starting point $(\boldsymbol{f}^{(0)}, \boldsymbol{h}^{(0)})$, since different initializations might lead to different limit point due to the nonconvexity of the objective function. In particular, since the first step of the alternating approach is a minimization on the image, a clever initialization for the PSF is crucial to address the sequence in the "basin of attraction" of the desired stationary point. While for $\boldsymbol{f}^{(0)}$ it is rather common to start with a constant image (and several tests with different choices showed very similar results), the initial PSF $\boldsymbol{h}^{(0)}$ has been chosen in two ways: the autocorrelation of the ideal PSF (possibly reiterated until the SR bound is satisfied) and the ideal PSF plus a small constant chosen to satisfy normalization and SR value.

As concerns the choice of the number of iterations, for the inner subproblems we used $N_f^{(k)} = 50$ and $N_h^{(k)} = 1$ for all k, while for the outer cycle we performed 3000 iterations. We remark that for the specific content of our images we do not perform an early stopping of the iterations to achieve a regularized solution since the minimizers of the KL divergence are sparse objects (*nightsky* [18] or *checker-board* [19] effect). However, we recall that our approach can be easily extended to incorporate regularization terms in the objective function, as a ℓ_1 constraint on the image (inducing sparsity) and/or a Tikhonov term on the PSF (inducing smoothness). In these cases, some Poisson-based criterion for the choice of the parameter(s) would be necessary (see e.g. [20, 21]).

3.3. Results

The accuracy of the reconstructions has been evaluated with different indices, namely the MARE for the image and the RMSE for the PSF, defined respectively as

$$MARE = \frac{1}{q}\sum_{i=1}^{q} \frac{|m_i - \tilde{m}_i|}{\tilde{m}_i} \quad ; \quad RMSE = \frac{\|\boldsymbol{h} - \tilde{\boldsymbol{h}}\|_2}{\|\tilde{\boldsymbol{h}}\|_2}, \tag{5}$$

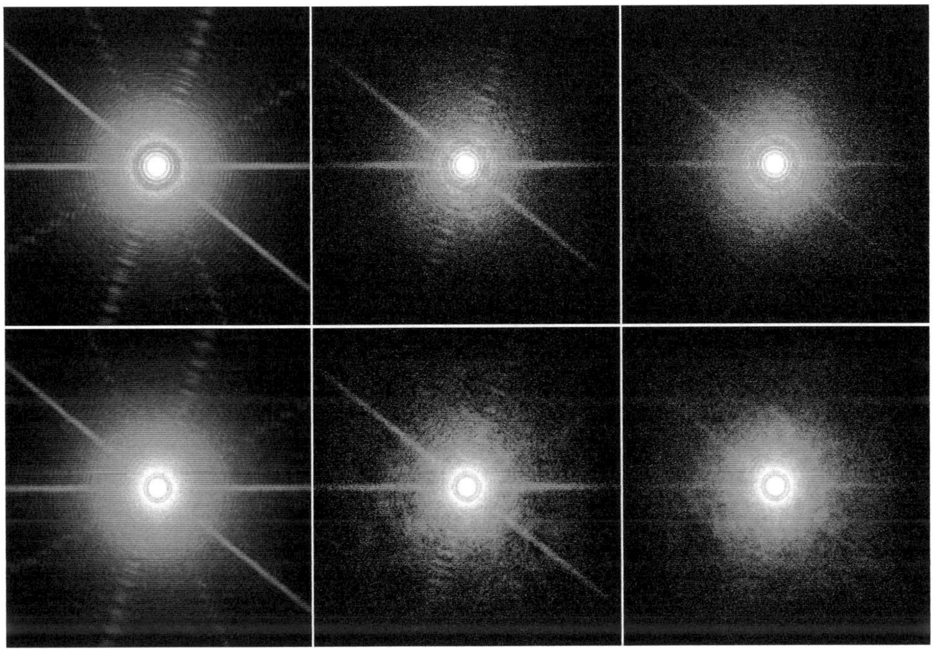

Figure 2. Left panels: true PSFs with SR = 0.81 (top row) and 0.60 (bottom row). Middle panels: reconstructed PSFs in the SC200 test when $\boldsymbol{h}^{(0)}$ is the autocorrelation of the ideal PSF. Right panels: reconstructed PSFs in the SC200 test when $\boldsymbol{h}^{(0)}$ is the ideal PSF plus a constant. All images are displayed in log scale.

where q is the number of stars, m_i and $\tilde{m}_i$ are the reconstructed and the true magnitudes, $\boldsymbol{h}$ and $\tilde{\boldsymbol{h}}$ are the reconstructed and true PSFs and $\|\cdot\|_2$ denotes the Euclidean norm.
The results are shown in table 1 while in figure 2 we reported the reconstructed PSFs for the SC200 dataset. In particular, we can notice that the algorithm is able to recover the stars with a very high accuracy, with values of the MARE which are always below 0.1%. At the same time, both the PSFs are also well recovered, with reconstruction errors varying around a few percent. As concerns the comparison between the two initializations, our results show that the autocorrelation of the ideal PSF seems to provide better reconstructions of the PSF for the higher SR, as attested both by the lower RMSE in table 1 and by the plots in figure 2, in which we can notice a better recovery of the PSF structures. No significant differences are present in all the three tests when SR = 0.60. We have to remark that we always observed a convergent behaviour in the objective function while, in few cases, the errors of the PSF are still decreasing after 3000 iterations, thus indicating that a larger number of iterations could still improve the solutions.

4. Conclusions and future work

In this paper we present a general optimization approach for the minimization of the KL divergence in the case of blind deconvolution of images corrupted by Poisson noise. The minimization scheme we analyzed exploits the separability of the constraints, providing subsequences which converge to stationary points. Moreover, it is particulary suited for the feasible sets arising in the case studied, made up by box constraints combined with one equality constraint. The application to the reconstruction of star cluster images blurred by LBT-like PSFs seems to confirm that the method is able to reconstruct both the target distribution and the unknown PSF with very high accuracy and paves the way to possible integrations into specific codes developed for accurate photometric and astrometric analysis of star clusters, as the so-called *StarFinder* [22]. Future work will involve also the generalization of the approach to multiple-image problems as the co-adding of images with different PSFs [23] or the blind deconvolution of multiple images provided by a Fizeau interferometer [24]. Moreover, effective strategies for the reduction of undesired boundary effects [25, 26] will be employed, in order to be able to reconstruct effectively clusters with stars positioned near the edges of the image.

Acknowledgments

This work has been partially supported by MIUR (Italian Ministry for University and Research), under the projects FIRB - Futuro in Ricerca 2012 (contract RBFR12M3AC) and PRIN 2012 (contract 2012MTE38N), and by INAF (National Institute for Astrophysics) under the project TECNO-INAF 2010 "Exploiting the adaptive power: a dedicated free software to optimize and maximize the scientific output of images from present and future adaptive optics facilities". The Italian GNCS - INdAM (Gruppo Nazionale per il Calcolo Scientifico - Istituto Nazionale di Alta Matematica) is also acknowledged.

References

[1] Bertero M and Boccacci P 1998 *Introduction to inverse problems in imaging* (Bristol: Institute of Physics Publishing)
[2] Hansen P C, Nagy J G and O'Leary D P 2006 *Deblurring Images: Matrices, Spectra and Filtering* (Philadelphia: SIAM)
[3] Csiszár I 1991 *Ann. Statist.* **19** 2032–2066
[4] Mahajan V N 1983 *J. Opt. Soc. Am.* **73** 860–861
[5] Prato M, La Camera A, Bonettini S and Bertero M 2013 *Inverse Probl.* **29** 065017
[6] Bonettini S 2011 *IMA J. Numer. Anal.* **31** 1431–1452
[7] Hadj S B, Blanc-Féraud L and Aubert G 2012 Space variant blind image restoration Rapport de recherche RR-8073 INRIA
[8] Hadj S B, Blanc-Féraud L, Aubert G and Engler G 2013 *ICASSP* (IEEE) pp 915–919
[9] Bonettini S, Zanella R and Zanni L 2009 *Inverse Probl.* **25** 015002
[10] Bonettini S and Prato M 2010 *Inverse Probl.* **26** 095001
[11] Bonettini S, Cornelio A and Prato M 2013 *SIAM J. Imaging Sci.* **6** 1736–1757
[12] Bonettini S and Prato M 2014 *Inverse Probl.* **30** 055004
[13] Zanella R, Boccacci P, Zanni L and Bertero M 2009 *Inverse Probl.* **25** 045010
[14] Zanella R, Zanghirati G, Cavicchioli R, Zanni L, Boccacci P, Bertero M and Vicidomini G 2013 *Sci. Rep.* **3** 2523
[15] Prato M, Cavicchioli R, Zanni L, Boccacci P and Bertero M 2012 *Astron. Astrophys.* **539** A133
[16] Esposito S, Riccardi A, Fini L, Puglisi A T, Pinna E, Xompero M, Briguglio R, Quirós-Pacheco F, Stefanini P, Guerra C J, Busoni L, Tozzi A, Pieralli F, Agapito G, Brusa-Zappellini G, Demers R, Brynnel J, Arcidiacono C and Salinari P 2010 *Adaptive Optics Systems II* (*Proc. SPIE* vol 7736) ed Ellerbroek L B, Hart M, Hubin N and Wizinowich P L p 773609
[17] Snyder D L, Helstrom C W, Lanterman A D, Faisal M and White R L 1995 *J. Opt. Soc. Am. A* **12** 272–283
[18] Barrett H H and Meyers K J 2003 *Foundations of image science* (Hoboken: Wiley and Sons)
[19] Natterer F and Wübbeling F 2001 *Mathematical methods in image reconstruction* 2nd ed (Philadelphia: SIAM)
[20] Bardsley J M and Goldes J 2009 *Inverse Probl.* **25** 095005
[21] Bertero M, Boccacci P, Talenti G, Zanella R and Zanni L 2010 *Inverse Probl.* **26** 105004
[22] Diolaiti E, Bendinelli O, Bonaccini D, Close L M, Currie D G and Parmeggiani G 2000 *Adaptive Optical Systems Technology* (*Proc. SPIE* vol 4007) ed Wizinowich P L pp 879–88
[23] Lucy L B and Hook R N 1992 *Astronomical Data Analysis Software and Systems I* (*A.S.P. Conference Series* vol 25) ed Worrall D M, Biemesderfer C and Barnes J pp 277–279
[24] Bertero M, Boccacci P, La Camera A, Olivieri C and Carbillet M 2011 *Inverse Probl.* **27** 113011
[25] Bertero M and Boccacci P 2005 *Astron. Astrophys.* **437** 369–374
[26] Anconelli B, Bertero M, Boccacci P, Carbillet M and Lantéri H 2006 *Astron. Astrophys.* **448** 1217–1224

Particle MCMC for Bayesian Microwave Control

P. Minvielle[1], **A. Todeschini**[2], **F. Caron**[3], **P. Del Moral**[4],

[1] CEA-CESTA, 33114 Le Barp, France
[2] INRIA Bordeaux Sud-Ouest, 351, cours de la Liberation, 33405 Talence Cedex, France
[3] University of Oxford, 1 South Parks Road, Oxford, UK
[4] UNSW, High Street, Kensington Sidney, Autralia

E-mail: pierre.minvielle@cea.fr

Abstract. We consider the problem of local radioelectric property estimation from global electromagnetic scattering measurements. This challenging ill-posed high dimensional inverse problem can be explored by intensive computations of a parallel Maxwell solver on a petaflopic supercomputer. Then, it is shown how Bayesian inference can be perfomed with a Particle Marginal Metropolis-Hastings (PMMH) approach, which includes a Rao-Blackwellised Sequential Monte Carlo algorithm with interacting Kalman filters. Material properties, including a multiple components "Debye relaxation"/"Lorenzian resonant" material model, are estimated; it is illustrated on synthetic data. Eventually, we propose different ways to deal with higher dimensional problems, from parallelization to the original introduction of efficient sequential data assimilation techniques, widely used in weather forecasting, oceanography, geophysics, etc.

1. Introduction

Unlike usual electromagnetic (EM) material characterization techniques [1], the microwave control problem involves to determine or check radioelectric properties (i.e. relative dielectric permittivity and magnetic permeability) of materials that are assembled and placed on the full-scaled object or system, from global scattering measurements (Radar Cross Section) [2].

An object or mock-up is illuminated by a monostatic radar that fulfills to a certain extent directivity and far-field conditions [3]. It illuminates the object at a given incidence with a quasi-planar monochromatic continuous wave (CW) of frequency f, the object backscatters a CW to the radar at the same frequency. With an appropriate instrumentation system (radar, network analyzers, etc.) and a calibration process, it is possible to measure the complex scattering coefficient. It sums up the EM scattering, indicating the wave change in amplitude and phase. It quantifies a global characteristic of the whole object-EM wave interaction in specific conditions (incidence, frequency, etc.). The scattering coefficient are measured for different transmitted and received polarizations. Eventually, various complex scattering coefficients $\mathcal{S}$ are measured at different wave frequencies ($f \in \{f_1, f_2, \cdots, f_{K_f}\}$, for K_f successive discrete frequencies) from a SFCW (Stepped Frequency Continuous Wave) burst, at different incidence angles ($\theta \in \{\theta_1, \theta_2, \cdots, \theta_{K_\theta}\}$, for K_θ different incidence angles) where the object is rotated with a motorized rotating support, at different transmitted/received linear polarizations (pol $\in \{HH, VV\}$). Otherwise, let assume that the objetc is axisymmetric and made of one metallic material, with its associated isotropic radioelectric properties weakly varying .The aim is to determine, from the global scattering measurement $\mathcal{M}$, the unknown isotropic local EM properties $(\epsilon_1, \mu_1), (\epsilon_2, \mu_2), \cdots, (\epsilon_N, \mu_N)$ along the object, where N is the number of different elementary areas.

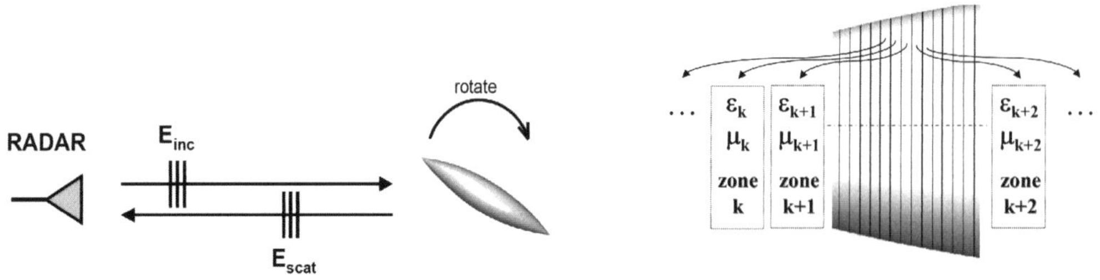

Figure 1. RCS measurement setup (left) - Elementary mesh zones (right)

It is shown in [2] that it can lead to a high dimensional inverse problem, that requires to go upstream a parallelized harmonic Maxwell solver (volume finite element/integral equation) [4]. Figure 2 sums up the entire inverse scattering problem. On the one hand, the RCS measurement process, that includes acquisition, signal processing, calibration, etc., provides the complex scattering measurement $\mathcal{M}$, with uncertainties. On the other hand, it would be useful to "row upstream" the Maxwell solver, in order to determine the unknown radioelectric properties, denoted by $\mathbf{x}$. Yet, even with recourse to HPC, there is no direct way to solve what turns out to be a high dimensional ill-posed inverse problem. On the contrary, the forward scattering model based on the resolution of Maxwell's equations can determine the scattering coefficients, given the EM properties, the object geometry and acquisition conditions (i.e. wave frequency, incidence, etc.). It lies in the resolution of Maxwell's equations, partial derivative equations that represent the electromagnetic scattering problem of an inhomogeneous obstacle. It is performed by an efficient parallelized harmonic Maxwell solver, an exact method that combines a volume finite element method and integral equation technique, taking benefit from the axisymmetrical geometry of the shape [4]. Discretization is known to lead to problems of very large sizes, especially when the frequency is high.

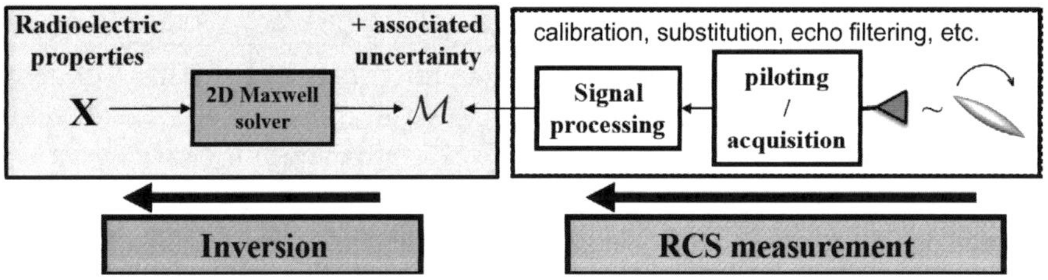

Figure 2. High dimensional inverse problem: "row upstream" the Maxwell solver

This current work is an extension to [2]. A Bayesian inference approach, based on "particle MCMC", is developed. It can perform estimation of material properties and determine a multiple components (Debye relaxation/Lorenzian resonant) material model. It provides various ways to deal with higher dimensional problems, from massively parallel computing to high-dimension oriented adaptations.

2. Problem statement : inference on a general HMM

The problem statement can be described as a general HMM (see figure 3). The graphical model is globally composed of a hierarchy of hidden states, at successive frequencies f_k, of a fixed hyperparameter $\mathbf{\Psi}$ (in blue) and of observations. Next, we detail the various items.

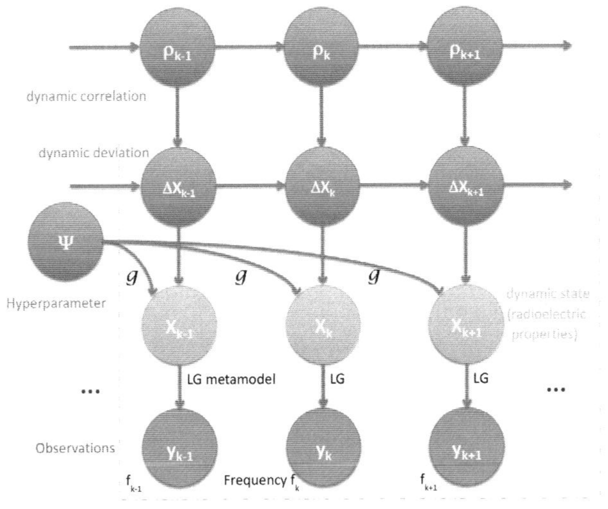

Figure 3. General Hidden Markov Model

State: The state $\mathbf{x_k} = \begin{bmatrix} \underline{\boldsymbol{\epsilon}}'_k & \underline{\boldsymbol{\epsilon}}''_k & \underline{\boldsymbol{\mu}}'_k & \underline{\boldsymbol{\mu}}''_k \end{bmatrix}^T$ iscomposed of the real ($'$) and imaginary ($''$) radioelectric components at frequency f_k (N elementary areas). It includes all the unknown EM properties that are to be estimated (at frequency f_k), i.e. the relative permittivity and permeability components of the N elementary zones (at frequency f_k). Omitting the indice k, the four components can be developed as: $\underline{\boldsymbol{\epsilon}}' = [\epsilon'_1 \cdots \epsilon'_N]^T$, $\underline{\boldsymbol{\epsilon}}'' = [\epsilon''_1 \cdots \epsilon''_N]^T$, $\underline{\boldsymbol{\mu}}' = [\mu'_1 \cdots \mu'_N]^T$ and $\underline{\boldsymbol{\mu}}'' = [\mu''_1 \cdots \mu''_N]^T$. Consequently, the state $\mathbf{x_k}$ is in a system space of dimension $4N$. Here, we are specially interested in the following decomposition:

$$\mathbf{x}_k = g(f_k, \boldsymbol{\Psi}) + \Delta\mathbf{x}_k \tag{1}$$

In this model, the EM properties $\mathbf{x_k}$ are supposed to partly follow a deterministic physical material model g and partly a random deviation term $\Delta\mathbf{x}_k$. The parametric material model $g(f_k, \boldsymbol{\Psi})$ depends on the frequency, where $\boldsymbol{\Psi}$ is the associated unknown hyperparameter. It is a sum of Debye relaxation/Lorenzian resonant terms (see [5] for details).The deviation (from model) $\Delta\mathbf{x}_k$ can be modeled as an AutoRegressive AR(1) model given frequential correlation ρ_k (modelled by a Markov process related to a random walk):

$$\Delta\mathbf{x}_1 \sim \mathcal{N}(\mathbf{0}, \mathbf{P}_1) \qquad \Delta\mathbf{x}_{k+1} = \mathbf{M}^{\rho}_k \cdot \Delta\mathbf{x}_k + \mathbf{w}_k \tag{2}$$

where $\mathbf{w}_k$: Gaussian noise ($\mathbb{E}(\mathbf{w}_k) \neq 0$). Notice that this stochastic process includes also spatial correlation (see [2] for details).

Observation: $\mathbf{y_k} = [\cdots]^T$ is composed of the complex scattering coefficients, measured at frequency f_k, for various rotation angles $\theta_1, \cdots, \theta_{K_\theta}$.

Likelihood model: The following likelihood model can be learned from intensive Maxwell solver computations on a petaflopic supercomputer at each frequency f_k ($\sim$ Born approximation, valid in a limited domain of interest [2]. It consists in the following Linear Gaussian metamodel:

$$\mathbf{y}_k = \left[\mathbf{A}_k \cdot \mathbf{x}_k + \mathbf{y}^0_k\right] + \mathbf{v}_k,\ \mathbf{v}_k \sim \mathcal{N}(0, \mathbf{R_k}) \tag{3}$$

The problem is to estimate jointly the fixed hyperparameter $\mathbf{\Psi}$, the dynamic states $\Delta\mathbf{x}_1, \cdots, \Delta\mathbf{x}_K$ and the dynamic frequency correlations $\rho_1, \cdots, \rho_K$, from the measurements $\mathbf{y_1}, \mathbf{y_2}, \cdots, \mathbf{y_K}$. It must be noticed that the general HMM contains a specific structure, i.e. the conditionally Linear Gaussian property given $\mathbf{\Psi}$ and $\rho_1, \cdots, \rho_K$.

3. Particle MCMC : PMMH (with Interacting KF)

3.1. Principle

Inference, in the general HMM context, can be performed by recent and powerful approaches, called "Particle MCMC" [6]. Since they involve two types of samplers, known as MCMC (Markov Chain Monte Carlo) and SMC (Sequential Monte Carlo), they are known to be computationally expensive. Here, we consider the specific method "Particle Marginal Metropolis-Hastings" (PMMH). In our context, it can solve the "smoothing" and parameter estimation simultaneously, managing to sample from the joint posterior distribution $p(\mathbf{\Psi}, \Delta\mathbf{x}_{1:K}, \rho_{1:K}|\mathbf{y_{1:K}})$. Is is based on a designed PMMH Markov chain which invariant distribution is the target joint distribution.

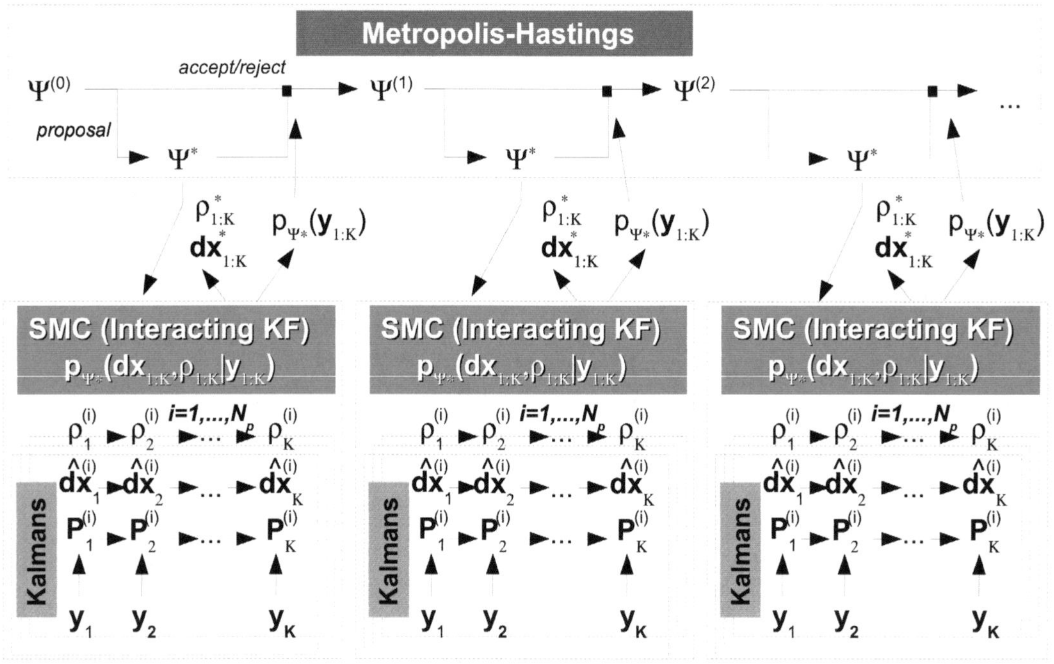

Figure 4. Rao-Blackwellised PMMH (with Interacting KF) scheme

Close to [7], our approach for the microwave control application lies on a PMMH multilevel stochastic algorithm. It is formed of 2 embedded levels (see figure 4):

- **MCMC higher level [Metropolis-Hastings]:** the Metropolis-Hastings methods is able to sample from $\mathbf{\Psi}^\star$.
- **SMC lower level [Rao-Blackwellized]:** A variance reduction strategy leads to a Rao-Blackwellised SMC method. It consists in a bank of interacting Kalman filters[8], that are able to compute the marginal likelihood $p(\mathbf{y_{1:K}}|\mathbf{\Psi}^\star)$ and sample from $(\Delta\mathbf{x}^\star_{1:K}, \rho^\star_{1:K})$.

3.2. Illustration (synthetic data)

Assumptions: The PMMH approach is applied on synthetic data, the dimension of which is relatively low. The object is composed of one material ($\dim(\mathbf{x_k})$=50 × 4), with spatial inhomogeneity. The associated material model (in frequency) is made of 2 terms (mentioned in figure 5): a Debye term (parameters: ϵ_∞ and ϵ_s) and Lorenzian term (parameters: f_r , μ_s and f_r). The deviation is simulated from the AR process model. Concerning the measurements, we consider $K = 20$ ($f \in$ [0.1-10 GHz]), $K_\theta = 100$ ($\theta \in [0^\circ - 180^\circ]$), $\dim(\mathbf{y_k})$=100 × 4.

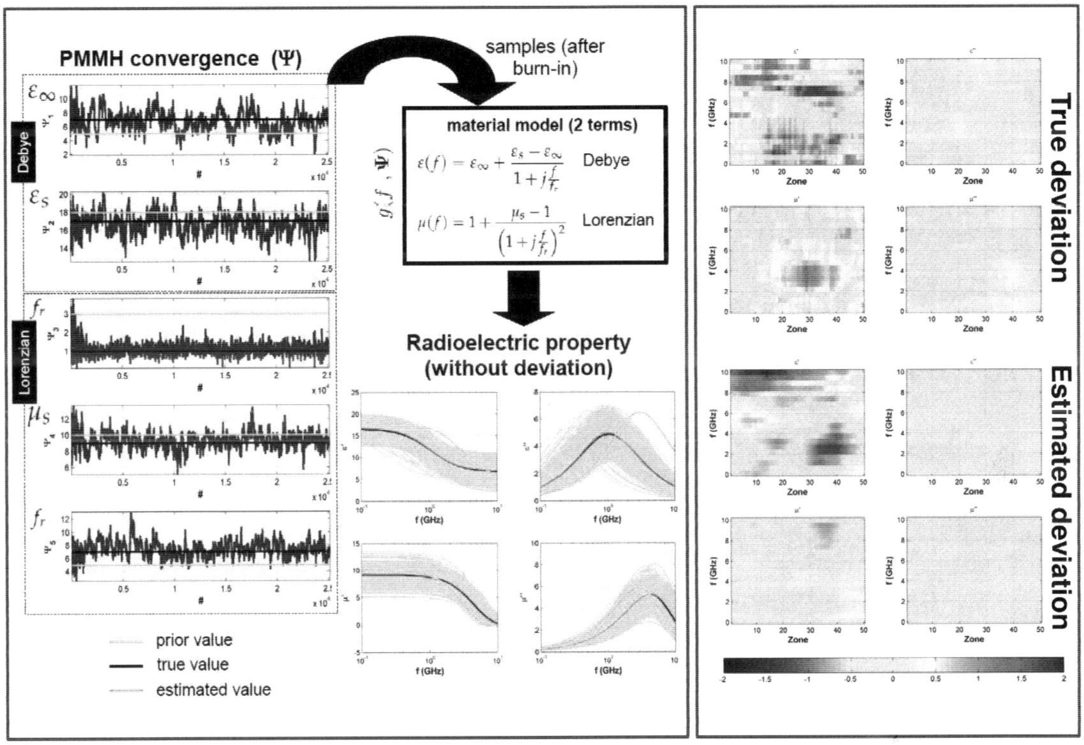

Figure 5. Results: chain convergence (left) - Deviations (right)

Inference process: For efficiency concerns, an adaptive PMMH algorithm [9] was developed. The MCMC level includes a tempering phase and a kernel mixture strategy. The SMC level is based on the classical SIR (Sampling Importance Resampling) algorithm, with one hundred "Kalman" particles. The whole PMMH behavior is illustrated in the left part figure 5. After a burn-in phase, the algorithm provides samples that approximate the posterior distribution. Propagated through the material model, the samples can be used to predict the deterministic part of the radioelectric properties. They are coherent with the true one. The right part of figure 5 compares the true deviation and the estimated one.It shows that the important deviations in μ' are detected while the insignificant ones can not be estimated.

3.3. Practical issues

The first practical issue turns out to be the high variation of marginal likelihood noisy estimates, due to a quite high dimension HMM (state dimension). It is solved by the adaptive strategy [9] and by tuning the SMC particle number (control the output noise on $p(\mathbf{y_{1:K}}|\Psi^\star)$ and

$(\Delta \mathbf{x}^\star_{1:K}, \rho^\star_{1:K})$), versus the MH step number, the proposal (control of the acceptance rate), etc. Le mention that another way could be to apply another related multilevel stochastic method, called SMC^2 [10]. And yet, the main practical issue is the expected computationally time expensiveness, $\sim$ 1 week on a standard PC despite the metamodeling speeding up. It will increase for higher state and observation dimensions. One way to overcome this issue is to turn towards massively parallel computing (at the SMC level), an important trend in Bayesian computational statistics. Another way hereinafter developed is to introduce high-dimension oriented adaptations and faster approximations.

4. High dimension adaptation: PMMH (with Interacting EnKF)

To accelerate PMMH inference, we have turned towards Ensemble Kalman Filter (EnKF) [11]. It is a Monte-Carlo alternative to (Extended) Kalman filter for huge dimensional state vector. It is widely used in sequential data assimilation: weather forecasting, oceanography, reservoir modelling, etc. It leads to a new original approach: a PMMH algorithm with Rao-Blackwellised SMC based on Interacting EnKF. The principle is to substitute Kalman filters by Ensemble Kalman filters with an efficient implementation, i.e. no empirical covariance manipulation and adaptation to a large number of data points (Sherman-Morrison-Woodbury formula). It provides similar results on dimension-limited problems, such as the above illustration. Concerning the complexity, the benefit is important for large state and observation dimensions (i.e. much higher than the herein problem). Notice that it is also compatible with massively parallel computing. The theoretical concern is that there is no longer unbiased estimate of unnormalized target density, required by PMMH to be an "exact approximation" of idealized MCMC (see [6]). Yet, in a practical viewpoint, the unbiased condition is actually unattainable, due to various bias sources, from RCS measurements (residual interfering echoes) to the approximate likelihood model.

5. Conclusion

A global Bayesian inference approach, consisting in a Particle Marginal Metropolis-Hastings (PMMH) that includes interacting Kalman filters, have been developed for micro-wave material control. It is based on two sampling levels, both MCMC and SMC. It simultaneously estimates model parameters and spatial/frequency deviations, at the cost of a high computational time that increases with dimensions. It can be reduced with massively parallel computing or/and inference approximations from sequential data assimilation.

References

[1] Knott E F, Shaeffer J F and Tuley M T 2004 *Radar cross section* (SciTech Publishing)
[2] Giraud F, Minvielle P and Del Moral P 2013 *Inverse Problems* **29** 095014
[3] Minvielle P, Tantar E, Tantar A and Berisset P 2011 *IEEE Transactions on Antennas and Propagation* **59** 2862–2871
[4] Stupfel B, Le Martret R, Bonnemason P and Scheurer B 1991 *Mathematical and numerical M* **2** 78–17
[5] Baker-Jarvis J, Geyer R G and Domich P D 1992 *Instrumentation and Measurement, IEEE Transactions on* **41** 646–652
[6] Andrieu C, Doucet A and Holenstein R 2010 *Journal of the Royal Statistical Society: Series B (Statistical Methodology)* **72** 269–342
[7] Nevat I, Peters G W, Doucet A and Yuan J 2010 *arXiv preprint arXiv:1006.3151*
[8] Doucet A, Godsill S and Andrieu C 2000 *Statistics and computing* **10** 197–208
[9] Peters G W, Hosack G R and Hayes K R 2010 *arXiv preprint arXiv:1005.2238*
[10] Chopin N, Jacob P E and Papaspiliopoulos O 2013 *Journal of the Royal Statistical Society: Series B (Statistical Methodology)* **75** 397–426
[11] Evensen G 2003 *Ocean dynamics* **53** 343–367

Optical imaging in a variational Bayesian framework

S Arhab*[1], H Ayasso[2], B Duchêne[1] and A Mohammad-Djafari[1]

[1] Laboratoire des Signaux et Systèmes (L2S, UMR8506: CNRS - SUPELEC - Univ Paris Sud)
[2] Grenoble Images Parole Signal Automatique (GIPSA-lab, UMR5216: CNRS - Grenoble-INP, Univ Joseph Fourier - Univ Stendhal)

E-mail: slimane.arhab@lss.supelec.fr

Abstract. We are interested in optical imaging of nano-structured man-made objects. Optical imaging is taken as a nonlinear inverse scattering problem where the goal is to retrieve the dielectric parameters of an unknown object. In addition to be nonlinear, such problems are also known to be ill-posed, which means that a regularization is required prior to their resolution. This is done by introducing *a priori* information which consists in the fact that object is known to be composed of compact homogeneous regions made of a finite number of different materials. This *a priori* knowledge is appropriately translated in a Bayesian framework by a Gauss-Markov-Potts prior. Hence, a Gauss-Markov random field is used to model the contrast distribution, whereas a hidden Potts-Markov field accounts for the compactness of the regions. The problem is then solved by means of a variational Bayesian approximation, which consists in approximating the joint posterior law of all unknown parameters in the Kullback-Leibler sense with a separable free form distribution. This leads to an implicit parametric optimization scheme which is solved iteratively. This inversion algorithm is applied to laboratory controlled experimental data.

1. Introduction

Optical Digital Tomographic Microscopy (ODTM) is a quantitative imaging technique [1] in which the opto-geometrical parameters of an object (shape, permittivity) can be determined with a good resolution from far field measurements. Thanks to the synthetic high numerical aperture obtained by varying the incidence angle, the lateral resolution achieved is at least twice better than that obtained with classical far field microscopes. The opto-geometrical parameters are determined by applying numerical inversion algorithm, with an important role played by the multiple scattering regime in terms of resolution. In classical optical microscopy, the latter is limited by the Rayleigh-Abbe criterion. However, when highly refractive objects are considered, resolutions beyond the Abbe-Rayleigh criterion can be achieved by accounting for multiple scattering through a rigorous resolution of Maxwell's equations. Optical microscopy is then taken as a non linear inverse scattering problem described through two coupled contrast source domain integral equations obtained by applying Green's theorem to the Helmholtz wave equations satisfied by the fields and by accounting for boundary and radiation conditions.

In the early 90's, several deterministic inversion algorithms [2, 3, 4] have been developed in order to deal with the nonlinear problem at hand, through an iterative minimization of a cost functional that accounts for both integral equations mentioned above and that expresses the discrepancy between the measured scattered fields and the fields associated to the best available solution (the sought contrast). However, in addition to be nonlinear, inverse scattering problems are also known to be ill-posed, which means that a regularization is required prior to their resolution. Such a regularization is usually done by introducing *a priori* information on the object, which is not easy with the latter techniques as this information must be introduced in the functional to be minimized. On the contrary, the probabilistic Bayesian framework allows us to take easily into account such information.

In the present case, we consider man-made objects that are known to be composed of compact homogeneous regions made of a finite number of different materials. This *a priori* is introduced through a Gauss-Markov-Potts model [5, 6], the marginal distribution of contrast being sought for as a Gaussian mixture [7] where each Gaussian law represents a class of materials, and the compactness of the regions being accounted for by a hidden Markov model.

In Bayesian methods the joint posterior distribution of the unknowns and parameters of the model is sought for by means of the Bayes rule. Generally, the estimators are computed in the maximum *a posteriori* (MAP) or posterior mean (PM) senses that are hardly tractable in the present case. Therefore, an approximation of the posterior law is introduced in order to obtain the desired solution at a reasonable computational cost. The latter is based upon the deterministic Variational Bayesian Approach (VBA, [9]) inspired by statistical physics [8], that approximates the posterior distribution by a separable density. As the calculations are then analytic, the rate of convergence of VBA is expected to be fast, at least much faster than that of the traditional MCMC approaches that are rather costly in terms of computation time as they attempt to estimate the posterior distribution by performing an empirical mean through a stochastic Gibbs sampling [10].

Such a method has been developed and successfully applied in the framework of microwave imaging [11]. It has also been successfully applied to optical imaging in a configuration different from that considered herein [12, 13] that leads to a different observation operator.

2. The experimental configuration and direct scattering problem

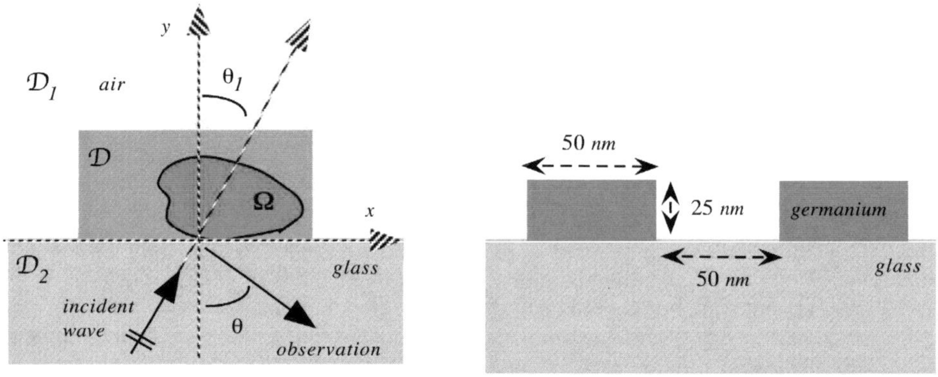

Figure 1. Experimental configuration.

The object considered herein, whose cross-section Ω is depicted in Figure 1, is made of two parallel 25-nm-height 50-nm-width germanium rods of long extent lying on a glass substrate and 50-nm distant from one other. The substrate is of known relative permittivity and its dimensions are large as compared to those of the rods, so that the configuration is modeled as follows: an object made of germanium lies in the upper layer of a stratified medium made of two semi-infinite half-spaces separated by a planar interface γ_{12}. The upper half-space $\mathcal{D}_1$ is air and the lower one $\mathcal{D}_2$ is glass. The different media are supposed to be lossless and they are characterized by their propagation constant k_m ($m = 1,\ 2$ or Ω) such that $k_m^2 = \omega^2\varepsilon_0\varepsilon_m\mu_0$, where ω is the angular frequency, ε_0 ($\varepsilon_0 = 8.854 \times 10^{-12}\,\mathrm{F\,m^{-1}}$) and μ_0 ($\mu_0 = 4\pi \times 10^{-7}\,\mathrm{H\,m^{-1}}$) are the dielectric permittivity and the magnetic permeability of free space, respectively, and ε_m is the relative dielectric permittivity of medium $\mathcal{D}_m$ ($\varepsilon_1 = 1,\ \varepsilon_2 = 2.25$ and $\varepsilon_\Omega = 29.8$). It can be noted that ε_Ω and Ω are supposed to be unknown in the inversion process.

The object is supposed to be contained in a test domain $\mathcal{D}$ ($\mathcal{D} \subset \mathcal{D}_1$) and we introduce a contrast function χ such that $\chi(\mathbf{r}) = (k^2(\mathbf{r}) - k_1^2)/k_1^2$, defined in $\mathcal{D}$ and null outside Ω. The object is illuminated by an incident wave generated by a Helium-Neon laser, operating at a 633 nm wavelength, coupled with a reflection microscope equipped with an interferometric device able to provide accurate measurements of the phase of the scattered fields [14]. The incident wave, whith an implicit time-dependence exp(-iωt), can be considered as a plane wave whose electric field E^{inc} is polarized in the γ_{12} interface plane along the axis of the rods so that a 2D-TM configuration is considered. The incident wave illuminates the rods from the glass side in direction θ_1 that can be varied in the range $\pm 50°$; hence $N_l = 9$ views are carried out at varying θ_1, each view being constituted of measurements of the scattered field in the far field domain $\mathcal{S}$ at $N_r = 683$ different observation angles θ in the range $\pm 54°$.

Modeling of the wave object interaction is done by means of two coupled domain integral equations obtained by applying Green's theorem to the Helmholtz wave equations satisfied by the fields and by accounting for boundary and radiation conditions. The first one, denoted as *observation* equation, relates the scattered electric field $y(\mathbf{r}')$ observed on a measurement domain $\mathcal{S}$ ($\mathbf{r}' \in \mathcal{S}$) to the Huygens type sources $w(\mathbf{r})$ induced within the object ($\mathbf{r} \in \mathcal{D}$) by the incident wave, i.e. $w(\mathbf{r}) = \chi(\mathbf{r})E(\mathbf{r})$ where $E(\mathbf{r})$ is the total field in the object, and involves Green's function of the stratified medium with source in $\mathcal{D}_1$ and observation in $\mathcal{D}_2$. The second one, the so-called *coupling* (or *state*) equation, relates the total field in the object to the induced sources $w(\mathbf{r})$ and the incident field $E^{inc}(\mathbf{r})$ and involves Green's function with both source and observation located in $\mathcal{D}_1$ [15, 16]. The forward problem then consists in first solving the coupling equation for E, knowing χ, then solving the observation equation for y knowing w. This is done from discrete counterparts of these equations obtained by means of a method of moments with pulse-basis and point matching. The test domain $\mathcal{D}$ is then partitioned into $N_{\mathcal{D}}$ elementary square pixels with side δ small enough to consider the field and the contrast as constant over each of them. Figure 2 displays the results obtained in this way compared to experimental data collected at Institut Fresnel in a controlled situation and a good agreement can be observed between both fields, once omitted the high level of noise in the experimental data that can be explained by the fact that the amplitude of the scattered field is very low as compared to that of the incident field.

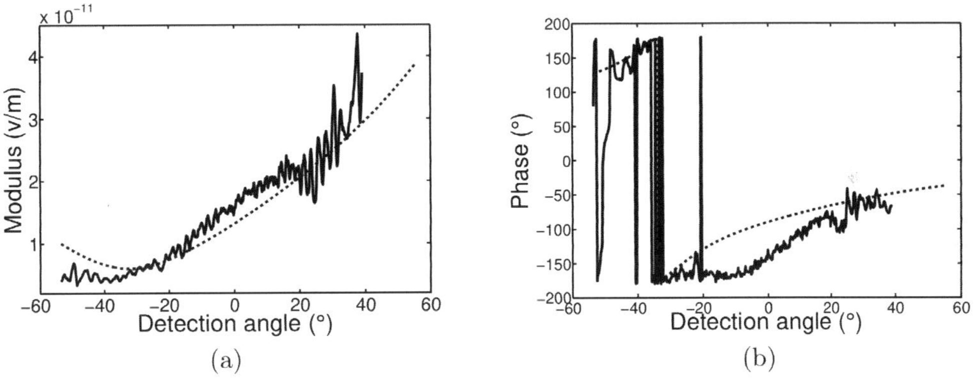

Figure 2. Modulus (a) and phase (b) of the calculated (dashed line) and measured (full line) scattered fields for the illumination $\theta_1 = 49°$.

3. The inverse scattering problem and results

Let us now introduce two variables, $\boldsymbol{\epsilon}$ and $\boldsymbol{\xi}$ that account for uncertainties and measurement and model errors and that are supposed to be centred and white and to satisfy Gaussian laws. The observation and coupling equations then read in a condensed form:

$$\begin{aligned} \mathbf{y}_l &= \mathcal{H}^{\mathcal{S}}\,\mathbf{w}_l + \boldsymbol{\epsilon}_l \\ \mathbf{w}_l &= \chi\,\mathbf{E}_l^{\mathrm{inc}} + \chi\,\mathcal{H}^{\mathcal{D}}\,\mathbf{w}_l + \boldsymbol{\xi}_l, \end{aligned} \tag{1}$$

where $\mathcal{H}^S$ and $\mathcal{H}^D$ are matrices whose elements result from the integration over the elementary pixels of the observation and coupling Green's functions, respectively, and subscript l accounts for the different views.

Let us now introduce *a priori* information that the object is composed of a finite number N_z of different materials. This is done by means of a hidden variable z_r ($z_r = z(r)$, $r = 1, ..., \mathrm{N}_r$) that associates a class of materials to each pixel r. Hence, when $z(r) = k$, the contrast value of pixel r is given by a Gaussian distribution $\mathcal{N}(\mu_k, \nu_k)$ with mean value μ_k and variance ν_k, the statistical parameters of the kth class. The object contrast χ is then defined by a multivariate Gaussian distribution that follows a conditionally independent Gauss mixture model:

$$\mathrm{p}(\chi|\mathbf{z}, \mu, \nu) = \prod_{k=1}^{\mathrm{N}_z} [1/(2\,\pi\,\nu_k)]^{\mathrm{N}_k/2} \, \exp\{-\frac{1}{2\,\nu_k} \sum_{r\in \mathrm{R}_k} [|\chi_r - \mu_k|^2]\}, \tag{2}$$

where $\mathrm{N}_k = \mathrm{card}(\mathrm{R}_k)$ is the number of pixels that belong to the kth class. It can be noted that pixels are conditionally independent between each other.

A priori information that the different materials are distributed in compact homogeneous regions is accounted for by means of a Potts-Markov model on $\mathbf{z}$:

$$\mathrm{p}(\mathbf{z}|\gamma) = (1/\Omega)\, \exp\{\frac{\gamma}{2} \sum_{r=1}^{\mathrm{N}_r} \sum_{r'\in \mathrm{O}(r)} \delta[z(r) - z(r')]\}, \tag{3}$$

where $\mathrm{O}(r)$ is a neighborhood of r, herein made of its four nearest neighbors, Ω is a normalization constant and γ determines the correlation between neighbors.

The conditional probability laws for scattered fields and induced sources are Gaussian distributions that account for the observation and coupling equations, respectively. They read:

$$\begin{aligned}
\mathrm{p}(\mathbf{y}|\mathbf{w}, \nu_\epsilon) &= [1/(2\pi\,\nu_\epsilon)]^{(\mathrm{N}_r\mathrm{N}_l)/2} \, \exp\{-\frac{1}{2\nu_\epsilon} \sum_{l=1}^{\mathrm{N}_l} ||\mathbf{y}_l - \mathcal{H}^S\,\mathbf{w}_l||^2_{\mathcal{S}}\} \\
\mathrm{p}(\mathbf{w}|\chi, \nu_\xi) &= [1/(2\pi\,\nu_\xi)]^{(\mathrm{N}_r\mathrm{N}_l)/2} \, \exp\{-\frac{1}{2\nu_\xi} \sum_{l=1}^{\mathrm{N}_l} ||\mathbf{w}_l - \chi\mathbf{E}_l^{\mathrm{inc}} - \chi\mathcal{H}^D\,\mathbf{w}_l||^2_{\mathcal{D}}\},
\end{aligned} \tag{4}$$

where $||\,.\,||_{\mathcal{A}}$ is the norm associated to the inner product in $\mathrm{L}_2(\mathcal{A})$, $\mathcal{A} = \mathcal{S}$ or $\mathcal{D}$.

From now on, all the hyper-parameters are grouped in the set θ ($\theta = \{\nu_\epsilon, \nu_\xi, \mu, \nu\}$) and conjugate probability laws are assigned to them. Hence $\mathrm{p}(\theta) = \mathrm{p}(\nu_\epsilon)\,\mathrm{p}(\nu_\xi)\,\mathrm{p}(\mu)\,\mathrm{p}(\nu)$, with $\mathrm{p}(\mu) = \prod_k \mathrm{p}(\mu_k)$ and $\mathrm{p}(\nu) = \prod_k \mathrm{p}(\nu_k)$, where the probability laws for the mean values μ_k are Gaussian distributions whereas that of the variances ν_k are inverse-gamma distributions as well as that of ν_ϵ and ν_ξ:

$$\begin{aligned}
&\mathrm{p}(\mu_k|\mu_o, \nu_o) = [1/(2\pi\,\nu_o)]^{1/2} \, \exp\{-\frac{1}{2\,\nu_o} \, |\mu_k - \mu_o|^2\} && k = 1, ..., \mathrm{N}_z \\
&\mathrm{p}(\nu_k|\alpha_o, \beta_o) = \mathcal{IG}(\alpha_o, \beta_o) = (1/\Gamma(\alpha_o))\, \beta_o^{\alpha_o}\, \nu_k^{-(\alpha_o+1)} \, \exp(-\beta_o\,/\nu_k) && k = 1, ..., \mathrm{N}_z \\
&\mathrm{p}(\nu_\tau|\alpha_\tau, \beta_\tau) = \mathcal{IG}(\alpha_\tau, \beta_\tau) = (1/\Gamma(\alpha_\tau))\, \beta_\tau^{\alpha_\tau}\, \nu_\tau^{-(\alpha_\tau+1)} \, \exp(-\beta_\tau\,/\nu_\tau) && \tau = \epsilon, \xi.
\end{aligned} \tag{5}$$

The meta-hyper-parameters $\mu_o, \nu_o, \alpha_o, \beta_o, \alpha_\tau$ and β_τ ($\tau = \epsilon, \xi$) are set to satisfy non-informative flat prior distributions.

It can be noted that an unsupervised context is considered here and the contrast, contrast sources and hyper-parameters of the model are estimated at the same time. The joint posterior distribution is then obtained by applying the Bayes rule:

$$\mathrm{p}(\mathbf{w}, \chi, \mathbf{z}, \theta|\mathbf{y}) \propto \mathrm{p}(\mathbf{y}|\mathbf{w}, \nu_\epsilon)\, \mathrm{p}(\mathbf{w}|\chi, \nu_\xi)\, \mathrm{p}(\chi|\mathbf{z}, \mu, \nu)\, \mathrm{p}(\mathbf{z}|\gamma)\, \mathrm{p}(\theta). \tag{6}$$

From this joint posterior law, different inferences can be done. Usually an estimator such as the maximum *a posteriori* (MAP) or the posterior mean (PM) is sought for. In the present case, these estimators are very difficult to obtain. This is the reason why a variational Bayesian method is adopted. The latter has been developed in order to perform approximate posterior inferences at a low computational cost compared to stochastic sampling methods [17, 18]. It consists in approximating the joint posterior law by a free form separable distribution where the unknowns are updated iteratively so as to minimize the Kullback-Leibler divergence:

$$\mathrm{KL}(\tilde{\mathrm{p}}(\Theta|\mathbf{f})||\mathrm{p}(\Theta|\mathbf{f})) = \int_\Theta \tilde{\mathrm{p}}(\Theta|\mathbf{f}) \log[\tilde{\mathrm{p}}(\Theta|\mathbf{f})/\mathrm{p}(\Theta|\mathbf{f})]\, d\Theta, \tag{7}$$

where $\Theta = \{\mathbf{w}, \chi, \mathbf{z}, \theta\}$. The free form distribution then reads [9]:

$$\begin{aligned} \tilde{\mathrm{p}}(\mathbf{w}, \chi, \mathbf{z}, \theta|\mathbf{y}) &= \prod_{l=1}^{\mathrm{N}_l} \prod_{r=1}^{\mathrm{N}_\mathcal{D}} \tilde{\mathrm{p}}(w_l(r)|\mathbf{y}) \prod_{r=1}^{\mathrm{N}_\mathcal{D}} \tilde{\mathrm{p}}(\chi(r)|\mathbf{y}) \prod_{r=1}^{\mathrm{N}_\mathcal{D}} \tilde{\mathrm{p}}(z(r)|\mathbf{y}) \prod_{k=1}^{\mathrm{N}_z} \tilde{\mathrm{p}}(\mu_k|\mathbf{y}) \\ &\times \prod_{k=1}^{\mathrm{N}_z} \tilde{\mathrm{p}}(\nu_k|\mathbf{y})\ \tilde{\mathrm{p}}(\nu_\epsilon|\mathbf{y})\ \tilde{\mathrm{p}}(\nu_\xi|\mathbf{y}), \end{aligned} \tag{8}$$

and is detailed in [13].

Figure 3 displays the results obtained by means of the above method, i.e. the behavior of the means and variances of the contrast of classes 1 and 2 during the iterative process and the contrast obtained after 600 iteration steps. In spite of the high level of noise that appears in the scattered field data, the algorithm succeeds in retrieving the two germanium rods, but with a better reconstruction for the left rod than for the right one, which can be explained by an unexpected dissymmetry in the noise level as it appears in figure 2.

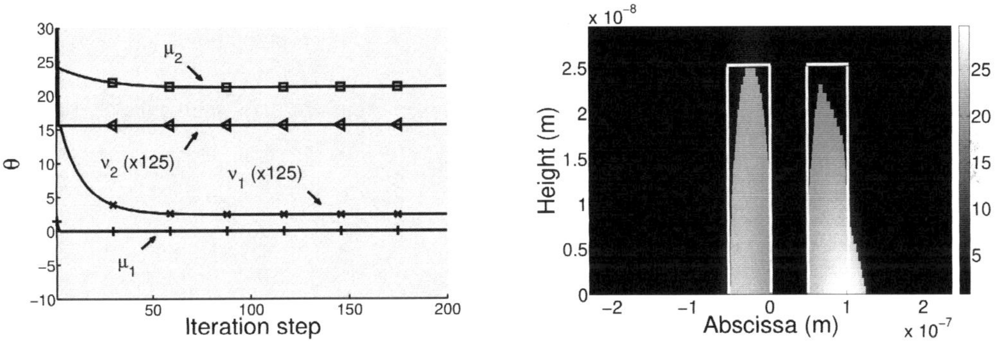

Figure 3. Behavior of the mean m_k and variance ν_k of the contrast for classes k=1 (air) and k=2 (germanium) during iterations (left) and the contrast retrieved by means of the variational Bayesian approach (right).

4. Conclusion

The results presented in this paper are a step toward the development of a new optical tomographic microscopy technique that yields a nanometric resolution far beyond that of standard optical microscopes limited by the Abbe-Rayleigh criterion. The dimensions of the object studied herein have been chosen in order to show that. Of course objects with smaller details could certainly be imaged. However, resolution limit achievable with this technique is not easy to quantify as it depends upon the level of noise that corrupts the data and upon the object itself. Indeed, the imaging technique is based upon a nonlinear inversion algorithm that benefits from multiple scattering due to the high dielectric contrast of the sought object and such a good resolution could certainly not be obtained with a lowly contrasted object imaged by means of a standard Born-based inversion method. On the other hand, deterministic nonlinear inversion methods such as the Newton-Kantorovich algorithm [2], the modified gradient method [3] or the contrast source inversion [4], are known to be limited in contrast level that can be accurately quantitatively retrieved, essentially due to the fact that they are based upon local minimization schemes. The method presented herein does not show such a drawback as it is not based upon such a strategy and as *a priori* information on the sought object introduced in the inversion algorithm allows us to reduce the eligible solution space. This can be quantified and constitutes the next step of this work, namely "what does *a priori* information add in terms of image quality?", as, from now on, it is more appropriate to quantify this quality rather than the standard resolution. Another interesting feature remains to be studied, i.e. the inversion of multifrequency data, that would certainly allows us to get a better image quality.

5. Acknowledgments

This work has been supported by Agence National de la Recherche under the convention n° ANR 12 BS03 006 02 (project "SURMITO"). The authors would like to acknowledge G. Maire, K. Belkebir and A. Santenac (Institut Fresnel, Marseille) for the experimental data.

6. References

[1] Lauer V 2002 New approach to optical diffraction tomography yielding a vector equation of diffraction tomography and a novel tomographic microscope *J. Microsc.* **205**(2) pp 165–176

[2] Joachimowicz , Pichot C and Hugonin JP 1991 Inverse scattering: an iterative numerical method for electromagnetic imaging *IEEE Trans. Antennas Propag.* **AP-39**(12) pp 1742–1752

[3] Kleinman RE and van den Berg PM 1992 A modified gradient method for two-dimensional problems in tomography *J. Comput. Appl. Math.* **42**(1) pp 17–35

[4] van den Berg PM and Kleinman RE 1997 A contrast source inversion method *Inverse Prob.* **13**(6) pp 1607–1620

[5] Tierney L 1994 Markov chains for exploring posterior distributions *The Annals of Statistics* **22**(4) pp 1701–1728

[6] Pieczynski W 2003 Markov models in image processing *Traitement du Signal* **20**(3) pp 255–278

[7] Féron O and Mohammad-Djafari A 2002 Image fusion and joint segmentation using an MCMC algorithm *J. Electron. Imaging* **14** 023014

[8] MacKay DJC 1995 Ensemble learning and evidence maximization *Proc. NIPS* 10.1.1.54.4083

[9] Smídl V and Quinn A 2005 *The Variational Bayes Method in Signal Processing* (New York: Springer)

[10] Robert C and Casella G 2000 *Monte Carlo Statistical Methods* (New York: Springer)

[11] Féron O, Duchêne B and Mohammad-Djafari A 2005 Microwave imaging of inhomogeneous objects made of a finite number of dielectric and conductive materials from experimental data *Inverse Prob.* **21**(6) pp S95–S115

[12] Ayasso H, Duchêne B and Mohammad-Djafari A 2010 Bayesian inversion for optical diffraction tomography *J. Mod. Opt.* **57**(9) pp 765–776

[13] Ayasso H, Duchêne B and Mohammad-Djafari A 2012 Optical diffraction tomography within a variational Bayesian framework *Inverse Prob. Sci. Eng.* **20**(1) pp 59–73

[14] Girard J, Maire G, Giovannini H, Talneau A, Belkebir K, Chaumet PC and Sentenac A 2010 Nanometric resolution using far-field optical tomographic microscopy in the multiple scattering regime *Phys. Rev.* A **82**(6), 061801

[15] Lesselier D and Duchêne B 1991 Buried, 2-D penetrable objects illuminated by line sources: FFT-based iterative computations of the anomalous field *Application of Conjugate Gradient Methods to Electromagnetics and Signal Analysis* (Progress In Electromagnetics Research vol 5) ed TK Sarkar (New-York: Elsevier) pp 351–389

[16] Greffet JJ 1989 Scattering of s-polarized electromagnetic waves by a 2D obstacle near an interface *Opt. Commun.* **72**(5) pp 274–278

[17] Beal MJ 2003 Variational algorithms for approximate Bayesian inference *University of London* PhD thesis

[18] Jaakkola TS 2001 Tutorial on variational approximation methods *Advanced Mean Field Methods: Theory and Practice* eds M Opper and D Saad (Cambridge: MIT Press) pp 129–159

Influence of partially known parameter on flaw characterization in Eddy Current Testing by using a random walk MCMC method based on metamodeling

Caifang Cai[1], **Thomas Rodet**[2] **and Marc Lambert**[1]

[1] L2S-SUPELEC, 3 Rue Joliot-Curie, 91192 Gif sur Yvette, France

[2] SATIE, ENS-Cachan, 61 Avenue du Président Wilson, 94230 Cachan, France

E-mail: caifang.cai@lss.supelec.fr, thomas.rodet@satie.ens-cachan.fr and marc.lambert@lss.supelec.fr

Abstract. First, we present the implementation of a random walk Metropolis-within-Gibbs (MWG) sampling method in flaw characterization based on a metamodeling method. The role of metamodeling is to reduce the computational time cost in Eddy Current Testing (ECT) forward model calculation. In such a way, the use of Markov Chain Monte Carlo (MCMC) methods becomes possible. Secondly, we analyze the influence of partially known parameters in Bayesian estimation. The objective is to evaluate the importance of providing more specific prior information. Simulation results show that even partially known information has great interest in providing more accurate flaw parameter estimations. The improvement ratio depends on the parameter dependence and the interest shows only when the provided information is specific enough.

Key words : flaw characterization, Eddy Current Testing, random walk, Markov Chain Monte Carlo (MCMC), metamodeling.

1. Introduction

In Eddy Current Testing (ECT), the objective is to estimate the parameters of the flaws present in the specimen examined, like their positions, dimensions and parameters related to their shapes, etc. In our situation, the processed data is the variation of impedance caused by the presence of a flaw.

So, if we denote $\boldsymbol{y} \in \mathbb{C}^M$ as the impedance variation where M is the measurment dimension, $\boldsymbol{x} \in \mathbb{R}^N$ as flaw parameters where N is the unknown parameter dimension, without considering the measurement noise, they can be related to each other by

$$\boldsymbol{y} = f(\boldsymbol{x}). \tag{1}$$

Here, $f(\boldsymbol{x})$ represents the physical phenomenon from $\boldsymbol{x}$ to $\boldsymbol{y}$. Obviously, it is not linear. The objective of flaw characterization is to estimate $\boldsymbol{x}$ from measurements of $\boldsymbol{y}$, of course noisy. One of the major difficulties in ECT is the numerical realization of the mathematical description of the physical phenomenon represented by $f(\boldsymbol{x})$. The method of moments [1, 2, 3] allows to get an accurate modeling of this phenomenon. However, it is computationally costly. This, in the inverse problem, limits the use of more sophisticated stochastic methods, like Maximum Likelihood (ML), Maximum A Posteriori (MAP) and Expected A Posteriori (EAP), since these methods are often solved by iterative numerical algorithms [4, 5] or Markov Chain Monte Carlo (MCMC) sampling methods [6, 7, 8]. Those require to calculate $f(\boldsymbol{x})$ many times.

A metamodeling method [9, 10, 11] provides surrogate options to get $f(\boldsymbol{x})$ with lower time cost. In metamodeling, this includes

- a well-trained database containing samples of $\boldsymbol{x}$ and the corresponding $f(\boldsymbol{x})$;
- an interpolator which allows to predict $f(\boldsymbol{x})$ for any input $\boldsymbol{x}$.

Bilicz *et al.* [12] present an *optimal* database training process by using the kriging interpolation. The kriging interpolation method [11, 13, 14] considers the interpolated samples as a Gaussian process guided by a covariance matrix. Interpolations minimizing the error distance have analytical expressions. Applying metamodeling in flaw characterization problem, the database only needs to be trained once at the beginning. When a forward calculation is required, it only needs to perform the kriging interpolation, which is much less costly computationally. For this reason, it makes the use of MCMC methods possible in ECT flaw characterization problem.

The interest of using MCMC sampling method here is first to get EAP estimations of flaw parameters, second, to calculate the variance of posterior distribution, third, to analyze the influence of a given parameter on other parameter estimations. Similar works [15, 16] have been conducted in view of parameter estimation. We are going to focus on the influence of partially known information. This is managed by the means of the MCMC sampling method.

2. Parameter estimation using random walk MCMC method

2.1. Likelihood with Gaussian noise model

Considering Gaussian noise present in the measurements, the forward model can be represented by

$$\boldsymbol{y} = f(\boldsymbol{x}) + \boldsymbol{\epsilon}, \text{ with } \boldsymbol{\epsilon} \sim \mathcal{N}(0, \Sigma). \tag{2}$$

$f(\boldsymbol{x})$ is the non-linear model describing the relation between $\boldsymbol{x}$ and $\boldsymbol{y}$. In our situation, it is replaced by the metamodel database and the *kriging* interpolator. $\Sigma = ||\boldsymbol{\sigma}_y||^2\mathbf{I}_M$ is the covariance matrix of the complex Gaussian noise. $\mathbf{I}_M$ is an identity matrix of size $M \times M$. The real and imaginary parts of $\boldsymbol{\epsilon}$ both follow Gaussian distributions. Their variances are $\sigma_R^2 = \sigma_I^2$. The standard deviation $\boldsymbol{\sigma}_y = \sigma_R + j\sigma_I$.

The likelihood of such a complex multivariate Gaussian model can then be given by

$$p(\boldsymbol{y}|\boldsymbol{x}) = \frac{1}{(\pi\sigma^2)^M} \exp\Big\{ -\frac{1}{\sigma^2}||\boldsymbol{y} - f(\boldsymbol{x})||^2 \Big\}, \tag{3}$$

where $\sigma^2 = ||\boldsymbol{\sigma}_y||^2$. In our situation, it is calculated based on the data Signal-to-Noise Ratio (SNR).

2.2. Uniform prior model

As for the parameters of the flaw, we choose to use the following uniform distribution

$$\boldsymbol{x} \sim \mathcal{U}(\boldsymbol{a}, \boldsymbol{b}) : \; p(\boldsymbol{x}) = \prod_{i=1}^{N} p_i(x_i), \; p_i(x_i) = \begin{cases} \frac{1}{b_i - a_i} & a_i \leq x_i \leq b_i \\ 0 & \text{else} \end{cases}, \; i = 1, 2, \cdots, N. \tag{4}$$

$\boldsymbol{a} = [a_1, a_2, \cdots, a_N]^T$ and $\boldsymbol{b} = [b_1, b_2, \cdots, b_N]^T$ denote the inferior and the superior borders for each parameter in the metamodel database. Denote $\mathbb{U}^\circ$ as the support for $\boldsymbol{x}$, it is bounded by the inferior and superior borders $\boldsymbol{a}$ and $\boldsymbol{b}$.

2.3. EAP estimation

Bayesian methods impose prior distributions about flaw parameters to estimate their values from noisy measurements. The EAP estimation calculates the expectation of the posterior distribution which can be described by

$$\hat{\boldsymbol{x}} = \int_{(\boldsymbol{x}\in\mathbb{U}^{\circ})} \boldsymbol{x}\ p(\boldsymbol{x}|\boldsymbol{y})\ \mathrm{d}\boldsymbol{x} = \int_{(\boldsymbol{x}\in\mathbb{U}^{\circ})} \boldsymbol{x}\ \frac{p(\boldsymbol{y}|\boldsymbol{x})p(\boldsymbol{x})}{p(\boldsymbol{y})}\ \mathrm{d}\boldsymbol{x}. \tag{5}$$

where $p(\boldsymbol{y}|\boldsymbol{x})$ is the likelihood given in Eq. (3). $p(\boldsymbol{x})$ is the prior distribution given in Eq. (4). $p(\boldsymbol{y})$ is the residual which is independent of $\boldsymbol{x}$.

2.4. Random walk MCMC

In order to get the estimations of the posterior distribution $p(\boldsymbol{x}|\boldsymbol{y})$, we choose to use the following random walk Metropolis-Within-Gibbs (MWG) method as described in Tab. 1. It is a combination of the random walk Hasting-Metropolis method [6, 17, 18] and the Gibbs alternating sampling strategy [7, 19].

Table 1: scheme of random walk Metropolis-within-Gibbs method. K is the sampling number.

1). Initiation: $\hat{\boldsymbol{x}}^{(k)}$, $k = 0$, $i = 1$.

2). Alternating candidature proposition: $\hat{\boldsymbol{x}}^{(p)} = \hat{\boldsymbol{x}}^{(k)}$,

  2.1, **alternating parameter** $\hat{x}_i^{(p)}$:

    generate randomly $s_i \sim \mathcal{U}(-0.5t_i, 0.5t_i)$, $\hat{x}_i^{(p)} = \hat{x}_i^{(k)} + s_i$.

    if $(i = N)$ $i = 1$, *else* $i++$;

  2.2, **verification:**

    if $(\hat{x}_i^{(p)} \in [a_i, b_i])$ go to **3** , *else* repeat 2.1 .

3). Decision:

  if $\left[r = \frac{p(\hat{\boldsymbol{x}}^{(p)}|\boldsymbol{y})}{p(\hat{\boldsymbol{x}}^{(k)}|\boldsymbol{y})} > 1\right]$ accept $\hat{\boldsymbol{x}}^{(p)}$, $\hat{\boldsymbol{x}}^{(k+1)} = \hat{\boldsymbol{x}}^{(p)}$,

  else

    3.1, generate $u \sim \mathcal{U}(0, 1)$;

    3.2, sub-decision: *if* $(u < r)$ accept $\hat{\boldsymbol{x}}^{(p)}$, $\hat{\boldsymbol{x}}^{(k+1)} = \hat{\boldsymbol{x}}^{(p)}$, *else* reject $\hat{\boldsymbol{x}}^{(p)}$, $\hat{\boldsymbol{x}}^{(k+1)} = \hat{\boldsymbol{x}}^{(k)}$.

  $k++$.

4). Repeat 2, 3 until $k = K$.

Once applied the sampling method, we get a set of samples $\{\hat{\boldsymbol{x}}^k\}_K$ for $k = 1, 2, \cdots, K$ following the posterior distribution $p(\boldsymbol{x}|\boldsymbol{y})$ after a period of warm-up (or burn-in) time. Then the EAP estimation given in Eq. (5) can be approximated by averaging over all the samples:

$$\hat{\boldsymbol{x}} = \int_{(\boldsymbol{x}\in\mathbb{U}^{\circ})} \boldsymbol{x}\ p(\boldsymbol{x}|\boldsymbol{y})\ \mathrm{d}\boldsymbol{x} \approx \frac{1}{K}\sum_{k=1}^{K} \hat{\boldsymbol{x}}^{(k)}. \tag{6}$$

To appraise the corresponding estimation uncertainty, the variance of the generated samples can be used. It is defined by

$$\boldsymbol{\sigma}_p^2 = D[\boldsymbol{x}] = E\Big[\big(\boldsymbol{x} - E[\boldsymbol{x}]\big)^2\Big] \tag{7}$$

It corresponds to the variances of the posterior distribution along different parameters. $E(\boldsymbol{x})$ can be replaced by its EAP estimation given in Eq. (6), then $\boldsymbol{\sigma}_p$ can be approximated by

$$\boldsymbol{\sigma}_p \approx \sqrt{\frac{1}{K-1}\sum_{k=1}^{K}\left(\hat{\boldsymbol{x}}^{(k)} - \hat{\boldsymbol{x}}\right)^2}. \tag{8}$$

3. Parameter estimation with partially known information

In certain situations, more specific prior information can be obtained. For example, the lift-off of the coil is often considered to be known while the width of a surface crack is often known to be small. The interest of studying the influence of a partially known parameter is to evaluate the importance of such kind of partial prior information, more importantly, the influence on other parameter estimations. Here, we define:

Partially known: *if the parameter support* $\boldsymbol{x} \in \mathbb{U}$ *is smaller than the metamodel database support* $\mathbb{U}^\circ$*, which means* $\mathbb{U} \subset \mathbb{U}^\circ$*, then the parameter* $\boldsymbol{x}$ *is said to be partially known.*

According to this definition, two situations might happen. One is that certain elements of $\boldsymbol{x}$ are totally known, the other is that the support for certain parameters are subsets of $\mathbb{U}^\circ$. In situations with partially known parameters, the same MWG method can be used to solve the estimation problem. It is only necessary to eliminate the known parameters from $\boldsymbol{x}$ or reset the inferior and superior borders to the corresponding borders in $\mathbb{U}$.

4. ECT example

4.1. Ferromagnetic configuration and data simulation

In order to analyze the influence of partially known parameters, we use a simple example as sketched in Fig. 1. An opening crack affects the surface of a conductive plate. The crack center is located at the origin O. We wish to characterize the dimension of the crack. There are three unknown parameters in our problem: flaw depth d, length l and width w.

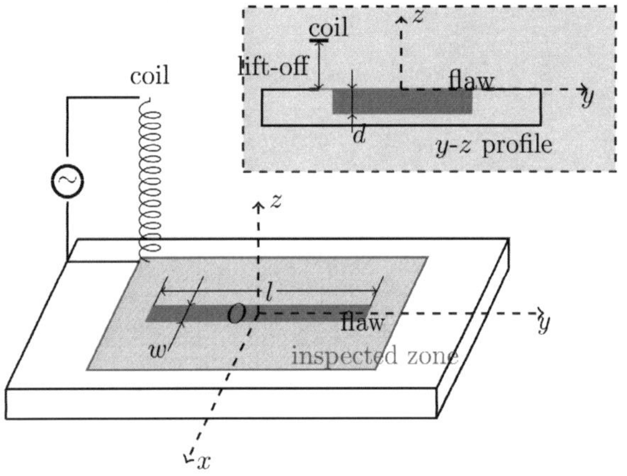

Figure 1: a 3D testing example: a flaw (in gray) at the surface of a conductive plate.

coil	
inner radius	1.0 mm
outer radius	1.75 mm
length	2.0 mm
number of turns	328
lift-off	0.303 mm
frequency	100 kHz
metal plate	
thickness	1.55 mm
conductivity	1.02 MS/m
inspected zone	
along x	$[-10, 10]$ mm
along y	$[-15, 15]$ mm
displacement step	0.5 mm

For the following analysis, the noise-free data is simulated by using the metamodel trained from simulated data using the method of moments implemented in CIVA [20]. Gaussian noise with SNR equal to 20 dB is then added numerically as described in Eq. (2). The testing sample is taken arbitrarily by using Latin Hypercube Sampling (LHS) method. Its parameter values are $d = 0.5428$ mm, $l = 8.7348$ mm and $w = 0.1523$ mm.

4.2. All unknown

First, we apply the Bayesian inverse method based on random walk MWG as described in Sec. 2 on the simulated data to estimate the three unknown parameters. Fig. 2 shows the MCMC samplings according to different profile views and the estimated results after eliminating warm-up samplings (around 700 out of 10^5 in total). In terms of CPU time, it takes around 4 minutes on a 3.4 GHz PC.

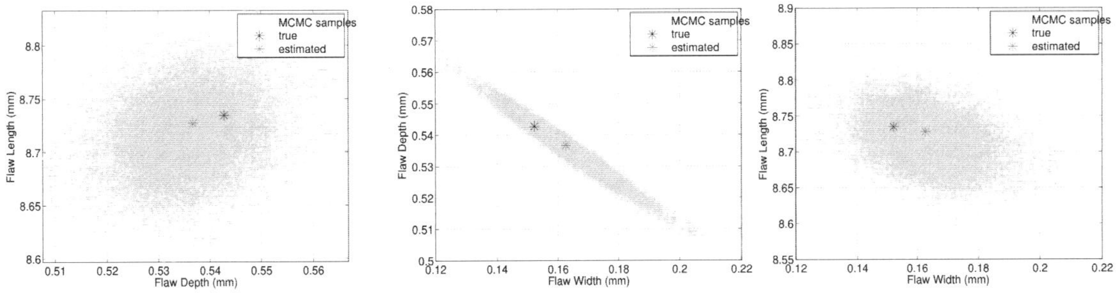

(a) d-l profile view. (b) w-d profile view. (c) w-l profile view.

Figure 2: MCMC samples ($K = 10^5$, walk step $\boldsymbol{t} = [0.01, 0.01, 0.04]$) of random walk MWG method for tested sample $d = 0.5428$ mm, $l = 8.7348$ mm, $w = 0.1523$ mm. The estimated results are $\hat{d} = 0.5385 \pm 0.0122$ mm, $\hat{l} = 8.7310 \pm 0.1739$ mm and $\hat{w} = 0.1597 \pm 0.0110$ mm.

We observe that the relative estimation error is less than 1.3% for flaw depth and length and around 4.9% for flaw width. From the w-d profile view, the FAP estimation is located in a narrow valley. This indicates that there is a determination difficulty between flaw width and depth. In other words, the flaw depth and width are highly dependent. In contrast, flaw length is less dependent on the other two parameters.

4.3. Partially known flaw width

In the above situation, we consider that all parameters are totally unknown. Strictly speaking, this is not correct, since we have benefited from prior information to get parameter inferior and superior borders in training the metamodeling database. But considering that the database support $\mathbb{U}^{\circ}$ often covers a much larger range than the one of our interest, so it can still be seen as *all unknown* situation.

Table 2: influence of partially known flaw width on estimations of the flaw depth and length.

Information level		**flaw depth**		**flaw length**	
		$\lvert\frac{\hat{\boldsymbol{x}}-\boldsymbol{x}}{\boldsymbol{x}}\rvert$ (%)	σ_p (mm)	$\lvert\frac{\hat{\boldsymbol{x}}-\boldsymbol{x}}{\boldsymbol{x}}\rvert$ (%)	σ_p (mm)
all unknown		0.7834	0.0122	0.0436	0.1739
partially known width	5	0.7864	0.0120	0.0426	0.1742
confidence radius r_c	2.5	**0.6777**	**0.0114**	**0.0434**	**0.1714**
(in times of $\hat{\sigma}_p = 0.011$ for flaw width	1.5	**0.2289**	**0.0111**	**0.0368**	**0.1685**
in *all-unknown* situation)	1	**0.1438**	**0.0099**	**0.0287**	**0.1644**
known width $w = 0.1523$		0.1114	0.0084	0.0261	0.1611

To study the influence of partially known parameters, we choose to give more prior information to the flaw width. As we observe from Fig. 2, the flaw width is highly dependent on the flaw depth but lightly on the flaw length. This makes it becomes a very good object to study influence of partial information. We define r_c as the radius of the confidence region for flaw width. It is the half distance from the inferior border to the superior border of $\mathbb{U}$ but given in ratio to the posterior standard deviation estimated in *all-unknown* situations. To simplify the simulation, $\mathbb{U}$ defined by r_c is always centered at the true value. So when $r_c = 3$, it means that more than 90% samples obtained in the *all-unknown* situation are located within the new support $\mathbb{U}$. Tab. 2 shows the estimation errors and standard deviations with partially known information.

We see that the estimation relative error and standard deviation start to decrease when r_c becomes less than 3. For flaw depth, an improvement of 85% on relative distance and 31% on standard deviation is achieved continuously while narrowing the confidence region. As for the flaw length, the improvement ratio is not as significant as for the depth. The improvement ratio is 40% for relative distance and 7% for standard deviation. This is because the flaw width is highly dependent on the flaw depth but less dependent on the flaw length, as seen from Fig. 2.

5. Conclusion and perspectives

This work analyzes the influence of partially known parameters on estimation of other unknown parameters. It is carried out by using a random walk MCMC sampling method based on a fast surrogate forward model. From the simulation results, we conclude that providing partially known information can help us to improve estimation accuracy. But this shows only when the confidence region for the parameter becomes smaller than 3 times of the range defined by the posterior standard deviation. In terms of improvement ratio, it depends on the dependence between parameters. The more they depend from each other the higher the improvement ratio becomes. For future work, it will be interesting to analyze the influence of nuisance parameters on flaw parameter estimation, for example, the lift-off of the coil which is often known roughly from manual measurement.

Acknowledgments

The French National Research Agency (ANR) has been providing financial support for this research work in the framework of ByPASS project. CEA-LIST is to be thanked for providing metamodeling databases.

References

[1] J R Bowler 1991 *Journal of Applied Physics* **70** 1107–14
[2] J R Bowler 1994 *Journal of Applied Physics* **75** 8128–37
[3] J Pavo and D Lesselier 2006 *IEEE Transactions on Magnetics* **42** 1419–22
[4] A P Dempster, N M Laird and D B Rubin 1977 *Journal of the Royal Statistical Society Series B (Methodological)* **39** 1–38
[5] J Nocedal and S J Wright 2006 *Numerical optimization* 2nd ed (Springer)
[6] W K Hastings 1970 *Biometrika* **57** 97—109
[7] L Tierney 1994 *The Annals of Statistics* **22** 1701–28
[8] M J Lombardi 2007 *Computational Statistics & Data Analysis* **51** 2688–700
[9] J K Sykulski 2008 *IET 7th International Conference on Computation in Electromagnetics* pp 44–49
[10] L Pronzato 2008 *Automatica* **44** 303 – 25
[11] R Douvenot, M Lambert and D Lesselier 2011 *IEEE Transactions on Magnetics* **47** 746–55
[12] S Bilicz, M Lambert and Sz Gyimthy 2010 *Inverse Problems* **26** 074012–27
[13] C K I Williams 1998 *Leaning in Graphical Models* **89** 599–621
[14] J P C Kleijnen and W C M Van Beers 2004 *The Journal of the Operational Research Society* **15** 303 – 25
[15] J I De La Rosa, G A Fleury, S E Osuna and M E Davoust 2006 *IEEE Transactions on Instrumentation and Measurement* **55** 112–22
[16] T Khan and P Ramuhalli 2008 *IEEE Transactions on Magnetics* **44** 1845–55
[17] N Metropolis, A W Rosenbluth, M N Rosenbluth, A H Teller and E Teller 1953 *The Journal of Chemical Physics* **21** 1087–92
[18] H Haario, E Saksman and J Tamminen 1999 *Computational Statistics* **14** 375–96
[19] S Geman and D Geman 1984 *IEEE Transactions on Pattern Analysis and Machine Intelligence* **PAMI-6** 721—41
[20] Miorelli R, Reboud C, Theodoulidis T, Poulakis N and Lesselier D 2013 *IEEE Transactions on Magnetics* **49** 2886–92

AUTHOR INDEX

Allain, P 25
Aquino, W 13
Arhab, S 43
Ayasso, H 43
Bonettini, S 31
Bonnet, M 13
Bonnet, S 7
Cai, C 49
Caron, F 37
Chamoin, L 19
David, O 7
Del Moral, P 37
Duchêne, B 43
Fouchard, A 7
Hervé, L 7
Kervrann, C 25
La Camera, A 31
Ladevèze, P 19
Lambert, M 49
Minvielle, P 37
Mohammad-Djafari, A 43
Peyrin, F 1
Prato, M 31
Rodet, T 49
Sixou, B 1
Todeschini, A 37
Waeytens, J 19
Wang, L 1